Erich Almendinger

Flora of Ann Arbor and vicinity

Erich Almendinger

Flora of Ann Arbor and vicinity

ISBN/EAN: 9783337268664

Printed in Europe, USA, Canada, Australia, Japan

Cover: Foto ©berggeist007 / pixelio.de

More available books at **www.hansebooks.com**

THE

CONSTITUTION AND BY-LAWS

OF THE

Ann Arbor Scientific Association

WITH

THE PROCEEDINGS

FOR THE

YEAR ENDING MAY 1, 1876.

Printed by Order of the Association.

ANN ARBOR:
COURIER STEAM PRINTING HOUSE.
1876.

Constitution and By-Laws.

CONSTITUTION.

ARTICLE I. NAME.

This Association shall be called the Ann Arbor Scientific Association.

ART. II. MEMBERSHIP.

All persons interested in any of the sciences, pure or applied, shall be deemed eligible to membership.

ART. III. OFFICERS.

The officers of this Association shall be a President, Vice-President, Secretary, Treasurer, and a Board of Censors.

ART. IV. DUTIES OF OFFICERS.

The duties of the officers of this Association shall be as laid down in the By-laws.

ART. V. ADOPTION OF BY-LAWS.

This Association shall have full power to adopt such By-laws as from time to time may be deemed necessary, by a majority vote of the members of the association.

BY-LAWS.

ARTICLE I. MEETINGS.

SECTION 1. This Association shall meet on the evening of the first Saturday of each month.

SEC. 2. Special meetings may be called by the President at his discretion, or on the written request of three members, due notice to be given as the President may direct.

SEC. 3. *Quorum.* Three members shall constitute a quorum for business at any regular meeting.

ART. II. MEMBERSHIP.

SECTION 1. Application for membership shall be made in writing, accompanied by a recommendation of at least one member of the Association, stating his or her residence and profession or occupation. Such application shall be referred to the Board of Censors, who shall, as soon as practicable, report thereon, either for or against the applicant.

SEC. 2. If the Board of Censors report favorably on any applicant, a two-thirds vote of all the members present at any regular meeting shall be necessary to election.

SEC. 3. If the Board of Censors report unfavorably, a two-thirds vote of all the members shall be necessary to election.

ART. III. DUTIES OF OFFICERS.

SECTION 1. *Duties of President.* The President shall preside at the meetings of the Association, when present, and shall perform such other duties as usually devolve on the presiding officer in deliberative assemblies.

SEC. 2. *Duties of Vice-President.* In the absence of the President, the Vice-President shall preside, and perform such duties as pertain to the office of President.

SEC. 3. In the absence of both President and Vice-President, the Association shall elect a President *pro tempore..*

SEC. 4. *Duties of Secretary. Clause* 1. The Secretary shall keep, in a book provided for the purpose, the Constitution, By-laws, Rules and Regulations of this Association, arranged for easy reference.

Clause 2. He shall also keep, in a book provided for the purpose, a correct record of the business transactions of this Association, and as full an abstract as possible of all papers, reports or discussions on scientific subjects.

Clause 3. He shall conduct the correspondence of the Association, and shall keep a file of all addresses, essays and other papers not otherwise provided for.

SEC. 5. *Duties of Treasurer.* The Treasurer shall collect and safely keep all money due the Association. He shall take possession of and hold all personal property belonging to this Association which is not otherwise provided for He shall keep a correct account of the same in a book provided for the purpose. He shall pay out money only on the order of the Secretary countersigned by the President. At the expiration of his term of office he shall make a full and correct report to the Association of the transactions of his office, and shall deliver up to his successor all property and papers of the Association in his hands.

If required, he shall give bonds for the faithful performance of his duty, in such sum and with such security as the Association shall deem proper.

SEC. 6. *Duties of the Board of Censors.* The Board of Censors shall examine all applications for admission, and report on the same to the Association.

It shall also be their duty to recommend to the President for appointment some member or members of the Association to read a paper on a volunteer subject or make a report on some selected subject,—one such paper, at least, to be read at each regular meeting. *Provided,* however, that nothing contained in this article shall be so construed as to prevent such paper or report being presented orally, and a synopsis of it presented in writing. The appointments shall be announced by the President two meetings in advance.

ART. IV. ELECTION OF OFFICERS.

SECTION 1. The officers of this Association shall be elected annually by ballot, and, except the Board of Censors, shall serve one year, or until their successors shall be elected.

SEC. 2. There shall be elected annually one member of the Board of Censors to serve three years, or until his or her successor is elected. In addition there shall be elected at a meeting in April, 1875, one member of said Board to serve two years, and one for the term of one year.

ART. V. FEES AND DUES.

The admission fee to this Association shall be *two dollars*, and the annual dues *one dollar*, payable annually in advance.

ART. VI. PENALTIES.

Any member in arrears more than one year for dues is thereby suspended without action on the part of the Association. If in arrears for more than three years, his name shall be stricken from the roll of membership.

ART. VII. DISCUSSIONS, ETC.

SECTION 1. Each member is requested to report to the Association at any meeting any fact or discovery of interest in the science in which he is most interested ; and to present for the examination of the Association any specimen, instrument or other object of special interest in such science.

SEC. 2. All papers presented to the Association shall be its property, to dispose of as it deems proper.

ART. VIII. AMENDMENTS.

Any proposed amendment to these By-Laws shall be offered in writing one month previous to action being taken thereon. Two-thirds of the members present at such regular meeting shall constitute a majority to decide on such amendment.

Any one or more of these By-Laws may be temporarily suspended by a majority of the members present.

———•••———

AMENDMENTS.

ART. IX. HONORARY AND CORRESPONDING MEMBERS.

SECTION 1. Any person prominent in science may be elected an Honorary Member by a unanimous vote at any regular meeting.

Any person in the active pursuit of science may be elected a Corresponding Member by a two-thirds vote of all members present at any regular meeting.

Resident members on removal from Ann Arbor or vicinity become Corresponding Members without action of the Association.

SEC. 2. No dues or other fees are required from Honorary or Corresponding Members.

Any communication from Corresponding Members will be referred to the Board of Censors, who shall report on it favorably before it is presented to the Association.

ART. X. ORDER OF BUSINESS.

1. Calling to Order.
2. Reading Minutes of last Meeting.
3. Applications for Membership received and referred.
4. Unfinished Business.
5. New Business.
6. Reports of Officers and Committees.
7. Balloting for Membership.
8. Papers and Discussions.
9. Reports and Display of Specimens and Apparatus.
10. Adjournment.

List of Members

Miss E. C. Allmendinger ---------------- ---May 1, 1875.
President J. B. Angell, LL. D----------- . --May 1, 1875.
Charles E. Beecher .... June 5, 1875.
Prof. W. W. Beman -------------------------- June 5, 1875.
Henry D. Bennett ------------------------- -March 4, 1876.
W. R. Birdsall ------------------- Nov. 6, 1875.
Rev. C. H. Brigham -------------------------- June 5, 1875.
Rev. F. T. Brown, D. D. ----------------- -Nov. 6, 1875.
Miss Lucy A. Chittenden ---------------- - Nov. 6, 1875.
Prof. J. A. Church ------------- --May 1, 1875.
Prof. H. N. Chute ----------------- April 17, 1875.
Miss Mary H. Clark* ------------- April 17, 1875.
Rev. Benj. F. Cocker, D. D., LL. D ------ April 17, 1875.
R. W. Corwin ------------------- April 17, 1875.
Miss Kate Crane, Ph. C. ---------------- April 17, 1875.
Mrs. Sallie A. Crane ------------------- Nov. 6, 1875.
Prof. J. B. Davis, C. E. ------------------ -April 17, 1875.
Miss Mary C. Douglas ------------- --------- June 5, 1875.
Prof. Silas H. Douglas, M. D. -------------- May 1, 1875.
Samuel T. Douglas, Ph. C. ---- -------------- May 1, 1875.
Prof. E. S. Dunster, M. D ------------------- -Nov. 6, 1875.
Miss A. E. P. Eastman ------------ ---------- May 1, 1875.
Ottmar Eberbach, --------------------------- April 17, 1875.
Mrs. B. C. Farrand ------------------------- -- Oct. 2, 1875.
Prof. C. L. Ford, M. D. ----------------- Oct. 2, 1875.
Mrs. M. E. Foster ---------------------- June 5, 1875.
C. George, M. D--- ------------------ --- May 1, 1875.
Prof. C. E. Greene, C. E. --------------------May 1, 1875.
G. G. Groff -- ----------------- ----April 17, 1875.
Israel Hall ----- ----------------------- -----Aug. 7, 1875.
Chas. N. B. Hall ------------------------------ Oct. 2, 1875.

------

*Deceased.

George Haller .... May 1, 1875.
W. D. Harriman...... . ...May 1, 1875.
Prof. M. W. Harrington............ ....April 10, 1875.
D. C. Hauxhurst, D. D. S............. ...... Nov. 6, 1875.
W. J. Herdman, M. D.................... ....Nov. 6, 1875.
W. H. Jackson, D. D. S. ....April 10, 1875.
O. C. Johnson............................ .....April 10, 1875.
C. J. Kintner....................... .. May 1, 1875.
Prof. J. W. Langley..... ... . .Oct. 2, 1875.
L. S. Lerch .... May 1, 1875.
Miss E. C. Merriam...... .... June 5, 1875.
Prof. John W. Morgan, M. D.. ... Nov. 6, 1875.
Prof. B. E. Nichols .... Nov. 6, 1875.
Prof. W. S. Perry.................. . Nov. 6, 1875.
Prof. W. H. Pettee........ ... . Oct. 2, 1875.
Prof. A. B. Prescott, M. D. . ... April 10, 1875.
Miss Louisa M. Reed ....June 5, 1875.
Henry W. Rogers.. ....Aug. 7, 1875.
Chas. Rominger, M. D... .. .. Nov. 6, 1875.
Prof. P. B. Rose, M. D.. .... April 10, 1875.
Miss C. A. Sager.......... ...... June 5, 1875.
J. Austin Scott .... April 1, 1875.
Ezra C. Seaman............. ... . June 5, 1875.
Wm. H. Smith................. .... Feb. 5, 1876.
Volney M. Spalding.................... ....Nov. 6, 1875.
Prof. J. B. Steere, Ph. D....................Feb. 5, 1876.
J. Taft, D. D. S. .... Oct. 2, 1875.
Prof. A. Ten Brook.................. .... June 5, 1875.
Charles Tripp..................... ....Jan. 8, 1876.
V. C. Vaughan ....April 10, 1875.
Miss Kate Watson....................... ....Aug. 7, 1875.
A. B. Wood... .... Jan. 8, 1876.
P. D. Woodruff...... .... May 1, 1875.

— •••• —

CORRESPONDING MEMBERS.

Prof. G. B. Merriman........................ .Albion, Mich.
G. W. Stone....... ....Albion, Mich.

PROCEEDINGS

OF THE

Ann. Arbor Scientific Association.

———•••———

SATURDAY EVENING, March 20, 1875.

Four persons met informally in Prof. Harrington's room, at the University, for the purpose of consultation with regard to the practicability of forming a scientific association or society.

After freely exchanging views with each other in regard to the desirability of such a society and the possibility of a failure, it was moved and carried that a committee of three, consisting of Prof. M. W. Harrington, Drs. W. H. Jackson and P. B. Rose, be appointed to confer with such persons as it might be thought were interested in such a society, and to report at a future meeting; also to report a plan of organization of such society if deemed practicable.

On motion the meeting adjourned to meet at the call of the committee.

———•••———

SATURDAY EVENING, April 10, 1875.

A meeting was called by order of the committee appointed at the meeting of March 20th. Present, seven persons.

The meeting was called to order by Prof. Harrington, chairman of the committee.

The committee reported as follows: That, from consultation with a number of persons, it was thought not only practicable but desirable that a Scientific Society should be formed in

Ann Arbor. The committee also reported a plan of organization, by submitting a draft of a Constitution and By-Laws.

The report of the committee was accepted, and on motion the Constitution and By-Laws were taken up article by article and adopted, as follows: (See Constitution and By-Laws.)

Owing to the small number present, it was thought best to adjourn for one week, before completing the organization and the election of officers. It was therefore moved and supported that when the Association adjourns, it be for one week; and that a committee of two be appointed by the chair, on the nomination of officers.

The motions were carried.

The chair then appointed P. B. Rose and W. H. Jackson such committee, after which the meeting adjourned to meet in one week.

——— •♦• ———

SATURDAY EVENING, April 17, 1875.

The meeting was called to order, eleven persons present.

The business in order was the report of the Committee on Nominations. The chairman, P. B. Rose, reported as follows:

President—Dr. B. F. Cocker.
Vice-President—Dr. A. B. Prescott.
Secretary—Prof. Merriman.
Treasurer—Dr. Jackson.
Board of Censors—Three years, Prof. M. W. Harrington; for two years, Miss Mary H. Clark; for one year, Prof. H. N. Chute.

On motion the report was accepted and the committee discharged.

The Association then proceeded with the election of officers by ballot, with the following result, Prof. Merriman having resigned the nomination of Secretary:

President—Dr. B. F. Cocker.
Vice-President—Dr. A. B. Prescott.
Secretary—Dr. P. B. Rose.
Treasurer—Dr. W. H. Jackson.
Board of Censors—Three years, Prof. Harrington; two years, Miss Clark; one year, Prof. Chute.

The President elect was then conducted to the chair, and addressed the Association with a few well-timed remarks, as follows :

That he appreciated the honor of election. Though devoted especially to metaphysical studies, he was in fullest sympathy with inductive science. The distinct observation and clear statement of a new fact in nature was a contribution to the understanding of the universe. Scientific truth is preëminently vitalizing to the mind. The best intellectual culture is scientific culture. He hoped they would not be content with mere organization. We are associated for work, for research. Let each feel the responsibility for success. Let each do his or her part, and let us not leave to this or the other person what we ought to do ourselves. Every member must be an active member. The failure of a Scientific Association in this university town, would be a disgrace, and he should be sorry to be known as a member of the failing concern.

No further business appearing, the Association adjourned.

P. B. ROSE, *Secretary.*

————•◦•————

MAY 1st, 1875.

The Association met and was called to order at 7½ P. M., Dr. Cocker, President, in the chair.

The minutes of the previous meeting were read and approved.

The following applications for membership were received and referred to the Board of Censors: President J. B. Angell, Prof. C. E. Greene, Prof. S. H. Douglas, Miss Kate Crane, Miss E. C. Allmendinger, Dr. A. Sager, Prof. J. A. Church, Miss A. E. P. Eastman, Geo. Haller, L. S. Lerch, W. D. Harriman, P. D. Woodruff, C. J. Kintner, S. T. Douglas, and Dr. C. George.

The Board of Censors having reported favorably on the above applications for membership, they were duly balloted for and declared unanimously elected.

The Secretary presented a bill for books and circulars for his office amounting to $9.05, which was referred to the Board of Censors.

Prof. Harrington was here introduced, and read a paper on the " Elevation of the Earth's Crust in Arctic Regions." (See Appendix A.)

A discussion of the facts presented and the results following them was participated in by the President, Dr. Cocker, and by Profs. Church, Davis, Greene, and Chute.

The Association then adjourned to the lecture-room of the Medical building, where Prof. Douglas exhibited the working of a new Magneto-Electric machine, known as the " Ladd Machine," the only one of the size in this country.

After the conclusion of the experiment, the Association adjourned.

P. B. ROSE, *Secretary.*

———•♦•———

JUNE 5TH, 1875.

The Association met and was called to order at 7½ P. M., Dr. Cocker, President, in the chair.

The following applications for membership were received : Profs. W. W. Beman and C. N. Jones, Miss L. M. Reed, Miss E. C. Merriam, Miss C. A. Sager, Dr. H. S. Cheever, Rev. C. H. Brigham, Profs. M. C. Tyler and A. Ten Brook, E. C. Seaman, Oscar Tucker, Miss Mary E. Douglas, Bryant Walker, Charles E. Beecher, Mrs. Mary E. Foster, and William H. Dopp.

The same were referred to the Board of Censors, who reported favorably, and on ballot they were declared legally elected.

The Board of Censors reported favorably on the bill of P. B. Rose, referred to them at the last meeting, and recommended its allowance and payment.

The report was accepted, and on motion a warrant was ordered drawn on the Treasurer for the amount of $9.05.

On motion a warrant was ordered drawn on the Treasurer for $2.50 for the bill of Fiske & Douglas.

It was then moved and supported that hereafter and until further notice, the time of meeting of the Association be changed to 8 P. M. Carried.

Prof. Harrington, in behalf of Bryant Walker and Charles E. Beecher, presented a paper comprising a list of the land and fresh-water shells found within a circuit of four miles about Ann Arbor, and collected by them. (See Appendix B.)

On motion, Messrs. Walker and Beecher were made a committee to make any additions and changes in the list which future observations should make necessary.

On motion, Miss Mary H. Clark and Miss E. C. Allmendinger were made a committee to make out a list of plants found growing within a radius of four miles of Ann Arbor.

Mr. R. W. Corwin was appointed a committee to make a list of all vertebrates, except fishes, found within the same radius.

Dr. W. H. Jackson was now introduced, and read a very interesting paper on "Hypertrophic Cement," illustrated by microscopic specimens and diagrams. He spoke first of the causes of hypertrophy in general, and then of hypertrophic cement under two heads :

First. The primary causes,—such as inflammation of the periosteum, death of the pulp, abnormal secretions of the mouth producing irritation of the gums or of the sentient terminal nerves in the dentine, irritation of the periosteum-dentine from undue pressure.

Second. He spoke of its histological structure and form as cap-shaped, laminated, modulated and penetrating.

There is no time of life between the formation of the temporary teeth and that of old age that is exempt from this hypertrophy. He referred to the fact that most histologists figure the cement as containing bone-corpuscles; and then stated that, from frequent and repeated observation of various specimens, he had concluded that true cement is nearly, if not entirely, destitute of what are called bone-corpuscles, and that the so-called bone-corpuscles found in cement are but the corpuscles which were developed when the cement-organ was in a state favorable to hypertrophy.

In hypertrophic cement, it is true, these corpuscles are very numerous, having broad, flattened bases, resting upon the external surfaces of the lamina, with canaliculi extending towards the periphery of the tooth. They vary much in size, and sometimes several are joined together. He considered them simply masses of periosteal tissue which have become enveloped within the rapidly growing cement.

There are occasional specimens when the hypertrophy is homogeneous. The corpuscles in such cases somewhat resemble bone-corpuscles. In these cases there are also present distinct tubuli, running parallel to each other with undulating or tortuous courses, somewhat resembling the dental tubules, but larger. These are not found in true cement. Again, that which is called the cement or bone-corpuscle is not necessary to the physiological structure of true cement, and, when found, is the result of a pathological condition of the part at the time of its formation; and, therefore, the presence of hypertrophic cement shows a pathological condition of the part, although the disturbance may not have been material.

A discussion of the facts set forth was engaged in by Profs. Prescott, Harrington, and Dr. Jackson. Prof. Harrington remarked that the shape of the corpuscles as described by Dr. Jackson was much like many seen in the bones of the ox and horse. He also said that many authors state that the bone-corpuscle is a hollow, hard structure which will be left behind after the bone has been dissolved away, and asked the essayist if he had observed such a fact in the cement. Dr. Jackson replied that he had never tried the experiment.

By a vote of the Association, Mr. L. V. Fletcher was requested to read a paper on Indian Mounds of Genesee county. Michigan. (See Appendix C.)

A discussion of the paper here followed.

On motion, Mr. Fletcher was requested to furnish the Association with a copy of his paper. (Many of the specimens described by Mr. Fletcher are deposited in the Museum of the University.)

Prof. Merriman called the attention of the Association to the report of the committee appointed by the Academy of Sciences of St. Petersburg to examine the merits of a new method of procuring the electric light. The committee, through its chairman, Mr. Wilde, reported so favorably that the Academy had awarded a medal to the inventor. The novel feature of the method consists in hermetically inclosing a slender rod of carbon in a glass cylinder from which oxygen is excluded, thus rendering combustion impossible. The carbon is thus made part of the electric circuit, and by its resistance becomes intensely incandescent, giving out a regular, uniform light without waste of carbon. The flicker and irregularity of the light, as commonly produced between two carbon points in air, which constitute a great objection to its use, are thus obviated.

Prof. Merriman was announced to read a paper on " Parhelia " at the August meeting.

On motion, the Association adjourned.

P. B. ROSE, *Secretary.*

JULY 3D, 1875.

The Association was called to order at 8 P. M., Dr. A. B. Prescott, Vice-President, in the chair.

The minutes of the previous meeting were read and approved.

The death of Miss Mary H. Clark, a member of the Board of Censors of this Association, was announced by the chair with appropriate remarks.

On motion of Prof. Harrington a committee of three were appointed by the chair to draft appropriate resolutions.

Profs. Harrington, Merriman, and Rose were appointed such committee, who reported the following, which were adopted :

IN MEMORIAM.

At a meeting of the Ann Arbor Scientific Association, held Saturday evening, July 3d, the following resolutions were adopted:

WHEREAS, By the interposition of Providence, one of our number has been removed by death from among us, Miss MARY H. CLARK; therefore

Resolved, That in her decease we have lost a valued member, one who took much interest in the founding of this Association, and who, by her scientific acquirements, her advice, and her hearty support, has contributed to its success.

Resolved, That we recognize in her removal a great loss to society, of which she was an esteemed member; to science, in which she always preserved a lively and intense interest; to the poor, who had learned to love her for her unremitting efforts in their behalf; and to the cause of education, in the pursuit of which she has spent forty years of unremitted labor.

Resolved, That we extend our heartfelt sympathy to her family and friends.

M. W. HARRINGTON,
G. B. MERRIMAN,
P. B. ROSE.

On motion, the Secretary was instructed to have a sufficient number of copies printed, and to present the same to the friends of the deceased.

2

On motion, Prof. Harrington was appointed a committee of one to procure the blanks required by the Association.

The death of Miss Clark having caused a vacancy in the Board of Censors, on motion the Association proceeded to an election to fill such vacancy, which resulted in the election of Miss C. A. Sager for the unexpired balance of the two years.

The Secretary presented the bill of R. A. Beal for printing, amounting to $1.50, and also one from J. Moore for envelopes, amounting to $1.00. The bills were referred to the Board of Censors, who reported favorably on them, and recommended their allowance and payment.

The report was received, and on motion adopted, and a warrant ordered drawn for the amounts.

The order of papers and discussions having been reached, Prof. Prescott read a paper on "The Aromatic Group of Organic Compounds; Their Significance in the Chemistry of Plants." (See Appendix D.)

A discussion of the facts presented by the paper followed.

Dr. Prescott reported to the Association the results of the analysis of four samples of Ann Arbor milk, as follows:

MILKMAN.	Specific grav.	Total solids.	Solids not fat.	Fat.
1. Van Giessen	1.0259	13.24	10.000	3.24
2. Ticknor	1.0300	13.71	10.327	3.38
3. Green	1.0280	11.89	10.120	1.77
4. Van Giessen	1.0290	13.41	10.109	3.30

Signed, N. G. O. COAD.

MAY, 1875.

This table shows that while each contained the proper amount of solids not fat, sample No. 3 was deficient in fat or cream, leaving us to conclude that it was largely made up of skimmed milk.

On motion, the Association adjourned.

P. B. ROSE, *Secretary.*

The Association met at the usual time and place, Dr. Prescott, Vice-President, in the chair.

The minutes of the preceding meeting were read and approved.

The names of the following persons were received for membership and referred to the Board of Censors: Israel Hall, Henry W. Rogers, Miss Kate Hale, Mrs. E. D. Kinne, Miss Kate Watson.

Prof. Merriman was here introduced, and read a paper on "*Halos and Parhelia*," of which the abstract follows:

The most usual forms presented by these phenomena are the following:

1. A colored circle or halo around the sun at a distance from it of 22°, and 2° to 3° in width, quite similar to a rainbow with the order of the colors reversed.

2. A second and similar circle about the first, at a distance of 46° from the sun.

3. An incomplete circle or inverted arc, also colored, tangent to the second halo at its highest point.

4. Two arcs tangent to the lower half of the second halo, and equally distant from its lowest point.

5. Short inverted arcs at the top and bottom of the inner halo.

6. A white band of light passing through the sun, parallel to the horizon, and not unfrequently extending quite around it.

7. Luminous spots, called *parhelia*, or, more commonly, "mock suns," where the horizontal band crosses the first halo, and sometimes, also, at other points on this band.

The same phenomena, but less brilliant, appear also about the moon.

The conditions necessary for these phenomena are : 1st, the presence of minute uniform crystals of frozen vapor in the higher strata of the air, forming a light cirrus cloud over the sun ; and 2d, comparative stillness of the atmosphere, that the positions of the crystals in falling may remain nearly uniform.

The most simple form of ice crystals is that of a hexagonal prism terminated by plane faces perpendicular to the sides. Suppose an immense number of such crystals, in every possible position, to be slowly falling in the air. Many of them will be in or near the position of minimum deviation relatively to the direction of the sun, and, unlike the others, will conspire to refract the sunlight in the same direction ; as a prism, when near the position of minimum deviation, can be rotated through a considerable angle without sensibly affecting the direction of the refracted ray. It is the combined action of the crystals in this position which produces the visible result. All those prisms, symmetrically situated with respect to the line through the eye and sun, and at a proper angular distance from it, will conspire to send the light to the eye, and the appearance produced is symmetrical with respect to that line, namely, colored circles with the sun as the center, the red (refracted least) marking the inner border, and the other colors following in the order of the spectrum. As the crystals within the circle can transmit to the eye no light at all, while those without and not in the position of least deviation may transmit some, the inner border is sharp and distinct, but the outer border fades into a feeble white.

The inclination of the lateral faces of a hexagonal prism is 60° and 120°; that of the ends with the sides is 90°. Taking the refractive index of ice as 1.31, it is easily shown that the mean deviation of light by passing through the angle of 60° is about 22°, and by passing through the angle of 90° is 46°, thus explaining the formation of the primary and secondary halos which are found to be respectively at those distances from the sun. The angle of 120° is too great for light to be transmitted.

The two halos described depend for their formation on the ice crystals having their axes at all angles. But as from their shape there would likely be an *excess* of crystals having their axes vertical or horizontal, distinct phenomena depending on these positions appear. If the sun is not too far above the horizon, the vertical prisms will throw an excess of light to the right and left, giving rise to lateral mock suns; the horizontal prisms will, in like manner, produce a mock sun above, and, if the sun's altitude is sufficient, also one below the real sun. These primary parhelia, when brilliant, may be the origin of secondary ones formed in the same manner.

Another effect of those crystals whose axes are vertical, produced by the light refracted through the terminal edges of 90°, is the inverted arc which touches the second halo at its upper point, and having its center at the zenith. The brightness which this arc frequently exhibits, and the order of its colors—violet within and red without, the red still being nearest the sun—give it a very close resemblance to an inverted rainbow high up in the sky. As this circumzenithal arc and the parhelia on the right and left of the sun are both due to the same position of the prismatic crystals, whenever one is visible the others generally are also, and this often in the absence of both the primary and secondary halos.

If the crystals assume a horizontal position, the same angles (90°) in like manner produce the two tangent arcs on the lower part of the secondary halo.

The preceding are all phenomena of refraction. But the light is also reflected from the surfaces of the prisms, as from a mirror. The vertical surfaces thus give rise to the white horizontal band passing through the sun, and, if many surfaces are oscillating about a horizontal position, they will occasion a like vertical band through the sun, just as the image of the sun or moon reflected from water not perfectly at rest is lengthened out into a vertical column of light.

Halos must be distinguished from coronæ, which are much smaller, appearing in fact quite close to the sun or moon, and having their colors in reverse order—the violet next the sun. The coronæ are due to the diffraction and interference of light caused by the small globules of water in the air. As the diminution in size of the coronæ indicates an increase in size of the watery spheres which cause them, this may be regarded as a token of approaching rain, which falls when the particles are no longer able on account of their size to float in the air. Halos are a less certain indication of a storm, though if their brightness is considerably obscured, they are not unfrequently followed by rain or snow.

The foregoing explanation of the cause of halos receives confirmation from the polariscope, which shows the light to be partially polarized in a plane tangent to the circle.

SYNOPSIS.

Positions of the Prismatic Crystals.

 I.   Prisms with axes at all angles.
 II. Prisms with axes vertical.
 III. Prisms with axes horizontal.

I.

1. Primary halo by angles of 60°.
2. Secondary halo by angles of 90°.

II.

1. Lateral parhelia (both primary and secondary) by angles of 60°.
2. Circumzenithal arc by angles of 90°.

III.

1. Tangent arcs and parhelia at upper and lower points of first halo, by angles of 60°.
2. Tangent arcs on the right and left of the lower half of second halo, by angles of 90°.

II., III.

Horizontal white band by reflection from vertical surfaces. Vertical white band by reflection from horizonal surfaces.

The paper was illustrated by black board drawings and preparations on glass. It was followed by a discussion engaged in by E. C. Seaman, Dr. Sager and Prof. Ten Brook.

Mr. J. D. Williams, of the Washtenaw County Pioneer Society, was introduced, and exhibited a supposed Indian relic in the form of a pipe-head with Egyptian peculiarities in features and arrangement of the hair. It was found about ten years ago on the surface of the ground at Boyden's Plains, eight miles from the city of Ann Arbor.

Dr. A. Sager gave the results of some " *Observations of the Development of some Dipterous Larvæ,*" as drawn from his note-book. On August 19, he found a group of some 15 or 20 jelly-like ovoid bodies as large as a pea, attached to each other by a common cord, on a small aquatic plant, which, when micro-scopically examined, were found to be composed of microscopic ova, invested with a glairy mucus, each mass containing from 2,000 to 3,000 eggs, curiously arranged in rows, the ova of each row being differently disposed. The enclosed embryos were distinctly visible through the transparent membranes. The em-bryos were so far developed as to exhibit the abdominal segments distinctly. Nearly in the center of the body was seen a dark, anteriorly bifurcate mass, which was composed of minute spheri-cal granules or probably cells, but not very distinct. There were twelve segments apparently, of the body, and the extrem-ity terminated with a pair of pincers. October 14th.—The viscera were now fully developed, the chain of nerve-ganglia, chiefly in pairs, the alimentary ˉcanal, the dorsal vessel and the respiratory tubes distinctly visible. The action of the dorsal vessel was beautifully exhibited. The structure exhibited this peculiarity, that instead of the usual form of valves, there were distinctly seen at short intervals, on the surface, opposite to each other, two or four small tubercles that completely closed the canal when in action, for an instant. The intestines were furnished with four long coveca, two ascending and two descending. The air vessels terminated in bifurcating processes on the last segment of the abdomen.

Prof. Harrington reported that Miss C. A. Sager and him-self had examined some twenty samples of tea obtained from

dealers in Ann Arbor, and found them free from adulterations
with other leaves.

On motion, the Association adjourned.

P. B. ROSE, Secretary.

OCTOBER 2, 1875.

The Association was called to order at 7½ P. M., Dr. A. B.
Prescott, Vice-President, in the chair.

The minutes of the previous meeting were read and approved.

The following proposals for membership were received in
due form and referred to the Board of Censors: Profs. C. L.
Ford, W. H. Pettee and J. W. Langley, of Ann Arbor; Prof.
J. Taft, of Cincinnati; C. N. B. Hall, and Mrs. B. C. Farrand.

The Board reported favorably on the above, and each being
duly balloted for, they were declared elected.

Prof. Harrington, from the committee on blanks, appointed
at the August meeting, reported the work completed, and the
necessary blanks obtained, with the exception of the warrant-
book.

The report was accepted, and the committee was authorized
to procure the warrant-book.

The bill of R. A. Beal for printing blanks, $6.00, was re-
ceived and referred to the Board of Censors.

Prof. Greene was here introduced, and read a paper on
"*The Removal of Obstructions under Water,*" of which the following is an abstract:

The improvement of navigable waters, as a public benefit, is
undertaken by the United States, and is carried on under the
Corps of Engineers. The expenses, compared with that of the
removal of similar materials on land, is very great in many cases,
as the work is often done at a disadvantage.

Soft materials may be removed by dredging. In very shoal
water, dredging may be done by hand, by means of a pole and

scoop. More commonly, in the depths of water to be found or made in channels for vessels, machines are employed. These may be classified as the scoop or dipper-dredge, the endless chain and bucket dredge, and the clam-shell dredge. A description of these forms, illustrated by drawings, was given. The last named is very effective. It will, for instance, remove the slabs and edgings which accumulate in rivers below saw-mills. The material dredged is emptied into scows, and towed to deep water or a suitable dumping ground.

The Engineer Corps designed a boat for use on the bars at the mouth of the Mississippi, which stirred up the mud by propellers. A ten-feet channel has been deepened to fifteen feet. The improvement was not permanent. Col. Eads is now trying the method of jetties or piers.

Boulders in shallow water may be removed by scows after a hole is drilled and an iron bar inserted and wedged. The scows are first lowered by letting in water, and then raised by bailing. In tidal waters a simple raft of logs may be employed, which lifts as the tide rises.

Blasting away ledges may be done in shallow water by drilling from a moored boat, and inserting a tin tube containing the charge, which may be fired by a water-proof fuse or by a battery. The space to be worked upon may be laid bare by a cofferdam. A steam drill is sometimes employed in ten or twelve feet of water, by placing the drill on the top of a strongly braced tripod to keep the steam cylinder from being chilled by contact with the water, and using a sufficiently long drill rod. Generally, in water of ten feet and over, the aid of divers is called in.

The dress of the diver, with his means of protection against cold, and the manner of supplying him with air, were then described in detail. The different sorts of blasts were described as surface, face and hole blasts. Gunpowder, dualin and other exploders were described, and an account given of the work on Blossom Rock in San Francisco Harbor, and on Hell Gate, New York.

The paper was quite fully discussed by different members of the Association. Prof. Ten Brook asked with regard to the use of a submarine boat or torpedo. Prof. Greene replied that the torpedo was quite successful, as it could be directed from the surface. Prof. Ten Book then related some experience he had had with a Mr. Bauer, of Germany, who had invented a submarine frigate, and was anxious to introduce it into this country during the late war.

Prof. Harrington gave the results of some investigations which he had recently made on some proprietary foods for babies, which are kept for sale by druggists. The first examined was *"Baby's Cereal Food."* He described it as a fine brownish powder, with a sweet, scorched taste. It was composed mostly of wheat-starch, altered by a wet heat. A little gluten and fragments of the envelopes of the wheat grain are present. The starch is scorched, and a little sugar is afterwards added.

2d. *"Ridge's Prepared Food."* It is a brownish, sweet powder, and is composed of well-bolted wheat-flour, scorched and a little sugar added.

3d. *"Sea-Moss Farine."* A violet-brown powder with a sea-water taste. It is composed of a small proportion of wheat-starch and of ground Irish moss (*Chondrus crispus*), with a few fragments of tissue not recognized, but supposed to be impurities in the Irish moss.

The Professor stated that he could not recommend any one of them as a food for babies, for at best they contain but little else than starch and sugar.

Dr. Jackson exhibited a specimen of rush in which the pods of the flower were changed to leaves.

Prof. Harrington exhibited two specimens of the Venus Fly-Trap in a living condition, obtained from Wilmington, North Carolina.

On motion, it was ordered that the Secretary be instructed to have printed the Constitution, By-Laws and list of members of the Association.

No further business appearing, the Association adjourned.

P. B. ROSE, *Secretary.*

A special meeting of the Association was held at 7½ o'clock, according to previous notice, and was called to order by the President, Dr. Cocker.

The minutes of the previous meeting were read and approved.

Bills of R. A. Beal, for printing, amounting to $3.75, were received and referred to the Board of Censors.

Prof. Harrington offered the following amendments to the By-laws, which were required to lie over one month, under the rule. (For these amendments, see By-Laws.)

It was moved and supported that a committee of three, consisting of the Board of Censors, be appointed to provide for a course of popular lectures before the Association, with full power to act in the selection of speakers and subjects. Carried.

Dr. Cocker read a paper on the "Nature of Life," written by Dr. Lionel S. Beale, F. R. S., of London. (See Appendix E.) The paper was very interesting.

In the discussion which followed, Prof. A. B. Prescott said that he had difficulty in obtaining a clear and consistent conception of the position of Dr. Beale, and of some other biologists, upon one point discussed in the very able paper of this evening. This point was as to the kind of force which produces and preserves the matter of living tissues, simply as matter, irrespective of structure or of life. Thus, in a piece of living nerve tissue, there are certain kinds of matter, made up of the elements carbon, hydrogen, nitrogen and oxygen. Certainly these elements are united by some sort of force or action ; and this is a transforming force or action (*i. e.*, it fills the chief definition of chemism) ; otherwise the matter would be only a mixture of dust and gases. The composition of these elements, to form certain kinds of matter is one thing ; the organization of this matter into certain outlines of structure is another thing; the vitalization of the matter would seem to be yet another thing. Mr. Prescott was predisposed, by all that he knew in science, to believe that chemical force is wholly distinct in operation from the force that produces organization, and from the force that effects vitalization. When chemical force has constructed the molecule, it has done. From its very nature, it can do no more. There is a current way of almost ascribing crystallization to chemical force; but of course no structure larger than the molecule can be due to chemical force. Now, it seemed almost self-evident, that the formation of *all* molecules is due to one sort of immediate cause, as much in tissues as in rocks, and as truly in the gelatinous substance of bone as in the calcium phosphate of bone. It appears to be the essential position of Dr. Beale, that chemical force cannot produce organization or vital action. Now, this strong position is not at all supported, but is weakened and controverted, by the doctrine that vital force forms molecules by uniting atoms. To suppose that vital or organizing force can hold together the elements in the substance albumen is but one degree less absurd than to suppose that chemical force can construct cells from albumen molecules. There is a gulf fixed between the formation of *matter*, homogeneous as it is under the most powerful microscope, and the *organization* of matter into cells; and this gulf can no more be crossed from the side of vitality than from the side of chemism. The main position of Dr. Beale, that chemical force does not effect organization or vital action, would not be affected in the least should it occur, in ten years or in fifty years, that albumen should be synthesized in the laboratory.

Mr. Prescott thought that organic chemists would not agree with Mr. Bloxam, as quoted, that no permanent constituent of tissue has been chemically synthesized. Perhaps it would be difficult to decide or to agree as to what are permanent or essential constituents of tissue; but perhaps it would be agreed that fats are such. Caproin and caprin are tolerably complex fats, and they have been synthesized.

Farther discussion by Mr. E. C. Seaman, Dr. Cocker and others, took place.

Mr. Prescott said he wished to add, in explanation of what he had already said, that he concedes that the organizing and vital forces in living bodies may and probably do *induce and modify* chemical actions in these bodies ; just as chemical action is affected by physical forces : iron and sulphur not uniting until a certain temperature is reached. But the union is chemical, in nature and in proportions, and, as we do not say that ferrous sulphide is a calorific compound, we should not say that albumen is a vital compound.

Miss Allmendinger exhibited an Indian pipe-bowl, very highly polished. It was plowed up on the farm of Mr. David Allmendinger, a few miles west of Ann Arbor.

The Board reported favorably on the bills of R. A. Beal, amounting to $10.25, and recommended their allowance and payment.

On motion, the report was adopted, and a warrant ordered drawn for the amount.

It was moved by Dr. Brigham, that the next regular meeting be held at 7 o'clock, on account of another lecture on the same evening. Carried.

On motion, the Association adjourned.

P. B. ROSE, *Secretary.*

———•◦•———

NOVEMBER 6, 1875.

The seventh regular meeting of the Association was held, Dr. Cocker in the chair.

The minutes of the special meeting were read and approved.

Applications for membership, properly recommended, were received from V. M. Spalding, W. R. Birdsall and D. C. Hauxhurst. They were referred to the Board of Censors, who reported favorably on them and the following additional names :

J. C. Watson, E. Olney, I. N. Elwood, W. J. Herdman, F. T. Brown, A. B. Palmer, B. E. Nichols, A. V. E. Young, Mrs. Sallie Crane, S. W. Smith, W. S. Perry, Miss L. A. Chittenden, C. Rominger, Miss O. W. Bates, S. A. Jones, E. S. Dunster, J. C. Morgan, D. M. Finley, F. A. Cady and F. H. Kimball.

On motion, Section 2 of Article 2 of the By-Laws was suspended for the evening, and the above candidates were elected *viva voce*.

Mr. Randall, photographer, of Detroit, presented the photographs of the following distinguished scientists and members of the American Association for the Advancement of Science: Prof. J. E. Hilgard, Washington, D. C.; Dr. J. L. Le Conte, Philadelphia; C. V. Riley, St. Louis, Mo.; and Prof. Edward S Morse, Salem, Mass.

On motion, a vote of thanks was extended to Mr. Randall for these photographs, and the Secretary was instructed to have them suitably framed.

On motion, the Association adjourned.

P. B. ROSE, *Secretary.*

————•••————

DECEMBER 4, 1875.

The Association met at the usual time and place.

In the absence of the Secretary, C. E. Greene was elected Secretary *pro tempore*.

On motion, Mr. E. C. Seaman obtained permission to read some remarks on the paper offered by Dr. Cocker on "Life" at the last meeting. (See Appendix F.)

Miss C. E. Allmendinger, from the Committee on the "Flora of Ann Arbor," made a final report, which was accepted. (See Appendix G.)

Mr. S. T. Douglas then read a paper on the "Colored Snow Fall of February, 1875." (See Appendix H.)

This was discussed by Prof. Langley, who thought that the dust might have been derived from Mt. Hecla of Iceland. Prof.

Harrington thought, from microscopical examination, that it might be the dust from the streets of Chicago or some other large city. Prof. Douglas expressed himself as convinced that the threads in the dust were Pele's hair. Dr. Jackson and others also partook in the discussion.

Prof. Harrington spoke of the tenacity of life in some land-snails, sent home from the Amazon by Prof. Steere. It is five years since they came here, yet last summer some of them awoke and moved about the case in which they were placed.

No further business appearing, the Association adjourned.

C. E. GREENE, *Secretary pro tem.*

———•◦•———

JANUARY 1, 1876.

The Association met at 7½ o'clock P. M., and, in the absence of both President and Vice-President, Prof. O. C. Johnson was elected President *pro tempore.*

The roll was called, and a quorum found present.

Prof. Harrington reported that Prof. Sill, of Detroit, would probably lecture before the Association in two weeks.

The Secretary reported the photographs framed, and presented the bill for the same, amounting to $7.25, which, on motion, was allowed, and a warrant ordered drawn for the amount.

The following bills were presented, and warrants ordered drawn for the amounts : C. G. Clark, for envelopes and stamps, $10.08 ; Prof. Harrington, for money paid janitors, etc., $5.75 ; H. C. Wilmot, posting bills, 38 cents.

The amendments to the By-Laws, offered at the regular meeting in November, 1875, were taken up article by article and adopted. (See Amendments, after Constitution and By-Laws.)

On motion, the Association adjourned, to meet Saturday evening, January 8, at seven o'clock.

P. B. ROSE, *Secretary.*

The Association met pursuant to previous adjournment, and was called to order by the President, Dr. Cocker.

The minutes of the meeting of January 1st were read and approved.

Propositions for membership of A. P. Wood and Chas. Tripp were received and referred to the Board of Censors.

Prof. Pettee was here introduced, and read a paper on " Barometric Measurements of Altitudes."

The object of this paper was to discuss questions relating to the degree of accuracy attainable in measuring the height of mountains by means of the mercurial barometer ; and to exhibit certain interesting results obtained by Mr. Pettee when engaged as assistant upon the State Geological Survey of California, under the direction of Prof. J. D. Whitney.

The detailed account of the experiments instituted in California, and of the manner of carrying on the work, may be found in the publications of that survey.

For a period of three years, barometric and thermometric observations were taken three times a day at three different stations, the altitudes of which above the sea-level were known from the spirit-level surveys of the Central Pacific Railroad. These observations were taken as the data from which to calculate in the usual way the differences of altitude between the respective stations. When the calculated differences of altitude were compared with the known differences, certain remarkable discrepancies became evident. The calculated differences were always higher in summer than in winter, higher at noon that at morning or night, and in some cases higher, in others lower, than the true differences. The practical benefits to be derived from such an investigation is the guide it affords to the explorer in selecting the time of day or the season of the year in which to make his observations for altitude, and in estimating the correction to be applied to calculating differences of altitude, if the observations have been made under unfavorable circumstances.

A discussion was engaged in by Drs. Brigham, Douglas and Morgan.

The Board of Censors reported favorably on Messrs. Tripp and Wood, and they were duly balloted for and elected.

Dr Herdman read a paper which called attention to some recent contributions to the knowledge of the state of Iceland, during the last year. (See Appendix J.)

On motion, the Association adjourned.

P. B. ROSE, *Secretary.*

● ● ● -- —

FEBRUARY 5, 1876.

In the absence of the President and Vice-President, Dr. C. H. Brigham was called to the chair.

The Secretary also being absent, W. J. Herdman was appointed Secretary *pro tempore.*

Dr. J. B. Steere, R. A. Beal and Wm. H. Smith were proposed for membership, and, on motion, the Secretary was instructed to cast the vote for their election.

A lecture was then delivered by Dr. Steere on the pottery, architecture, etc., ancient and modern, of the Amazon, and Peru. (See Appendix K.)

A plaster cast of a relic taken from an ancient mound near Rockford, Illinois, was presented to the Association by Mr. F. H. Kimball.

The thanks of the Association were tendered to Mr. Kimball, for the gift, and to Dr. Steere, for his interesting and instructive address.

A few remarks succeeded, on the viability of seeds, during which a statement was made by Prof. Harrington, in answer to an inquiry, that the British Agricultural Commission rarely suc-

3

ceeded, after thousands of experiments, in prolonging the vitality of seeds beyond fifteen years.

On motion, the Association adjourned until the next regular meeting.

W. J. HERDMAN, *Secretary pro tem.*

— ••• — —

The Association met in the Medical Lecture-Room, and was called to order by the President, Dr. Cocker.

The minutes of the previous meeting were read and approved.

Propositions for membership from H. D. Bennett and J. McDonald were received, and referred to the Board of Censors.

The Board reported favorably, and, on motion, the Secretary was authorized to cast the ballot for the Association, which was done in favor of their election.

Prof. Harrington, of the Board of Censors, gave notice of the following lectures and papers to be presented to the Association :

Public lecture for the middle of March, Prof. Langley, on the " Physical Theory of Hearing."

At the regular meeting in April, a paper by Mr. A. Macy, of Detroit, on " Iceland."

A paper, at the regular meeting in May, by Prof. J. B. Davis ; subject not yet known.

Paper, for the regular meeting in June, by Dr. C. George, on the " The Connection between Organic Germs and Disease."

Dr. Dunster was here introduced, and delivered a very interesting lecture on the " History of the Theory of Spontaneous Generation." (See Appendix L.)

A discussion of the subject was participated in by Mr. Seaman, Prof. Langley and Dr. Dunster, after which the Association adjourned.

P. B. ROSE, *Secretary.*

In the absence of the President, the Vice-President, Dr. Prescott, called the Association to order at 7½ o'clock.

The minutes of the preceding meeting were read and approved.

The following propositions for membership were received and referred to the Board of Censors, who reported favorably on the same: J. Austin Scott, Miss Eliza Ladd, Miss Annie Ladd.

They were duly balloted for and declared elected, V. C. Vaughan and V. M. Spalding acting as tellers.

Prof. G. B. Merriman and G. W. Stone, of Albion, Mich., were proposed as Corresponding Members, and, on motion, were declared elected.

This being the annual meeting, and therefore the night for the election of officers, on motion, the chair appointed Profs. Chute, Harrington and Greene to nominate officers for the ensuing year. They reported the following nominations:

*President—*Dr. A. B. Prescott.
*Vice-President—*Prof. C. E. Greene.
*Secretary—*W. D. Harriman.
*Treasurer—*C. Tripp.
*Member of Board of Censors for Three Years—*Prof. H. N. Chute.

The Association then proceeded to the election, V. M. Spalding and V. C. Vaughan again acting as tellers. Mr. Tripp, the nominee for Treasurer, having declined, Dr. Jackson was nominated in his place. The result of the election was as follows:

*President—*Dr. A. B. Prescott.
*Vice-President—*Prof. C. E. Greene.
*Secretary—*W. D. Harriman.
*Treasurer—*C. Tripp.
*Member of Board of Censors—*Prof. H. N. Chute.

Prof. Harrington, of the Board of Censors, spoke of the desirability of having the proceedings of the Association during the past year, as well as the papers read before it, and its Con-stitution and By-Laws, printed.

After some discussion of the subject, Prof. Ten Brook moved that the whole matter be referred to a committee of three, to be appointed by the chair.

An amendment was then proposed that the whole matter be referred to the Board of Censors, with power to proceed with the publication if thought desirable. This was accepted by the mover of the original motion, and, on being put to vote, the motion thus amended was carried.

Bills for printing and bill-posting were presented and re-ferred to the Board of Censors, who reported as follows:

R. A. Beal, printing $9 00
H. C. Wilmot, bill-posting 75
M. W. Harrington, janitor fees, and Mr. Macy's traveling expenses 8 75

Mr. A. Macy, of Detroit, was here introduced, and read an interesting paper on "Iceland," giving a history of the country and its inhabitants.

Mr. V. M. Spalding gave the results of some of his investi-gations in the Embryology of the Chicken. He also related his observations on the "Migrations of Chlorophyll-grains." (See Appendix M.) Both papers were illustrated by blackboard sketches.

On motion, a vote of thanks was tendered Mr. Macy for his very pleasing and interesting lecture. The motion was adopted by a rising vote.

No further business appearing, on motion the Association adjourned.

P. B. ROSE, *Secretary.*

NOTE.—The following public and advertised lectures were given to the Association and the public at other than the regular meetings:

Prof. J. Watson, on the "*Transit of Venus.*"

Prof. C. L. Ford, on the "*Anterior Extremity, Human and Comparative.*"

Prof. J. W. Langley, on the "*Physical Theory of Hearing.*"

APPENDIX.

CONTAINING IN FULL MANY OF THE PAPERS
AND LECTURES BEFORE THE ANN ARBOR
SCIENTIFIC ASSOCIATION.

A.

THE ELEVATION OF ARCTIC REGIONS.

During my year's stay in the Territory of Alaska, I picked up some evidence of the gradual elevation of the land going on at this time. I will first give that seen by myself, and then refer to the proofs of upheaval, of which I was told while there, or which I have found already recorded.

Amaknak Island is about two miles long by one broad, and lies in Captain's Harbor, a deep indentation in the northern end of Unalaska, one of the largest of the Aleutian Islands. Amaknak is composed of three or four distinct masses of hills or ridges connected by stretches of lowlands which are nearly level, and not more than 30 feet above the level of high water. The southernmost of these level stretches shows a distinct and fine series of elevated beaches. They are six in number, and from east to west each one is rather higher than the one next to it. The easternmost one is quite short, extending from a point of rocks on the northern wall. It reaches out to the quiet water on the eastern side of the neck. It is about 10 feet higher than the next, a much greater difference than between any other two successive beaches. When this beach was formed, the hill-masses between which it lies were distinct islands. It was not until the fourth beach was formed that the strait was entirely closed.

The east side of the lowland just described borders a land-locked passage of water. It shelves gradually down to the water's edge. There is never more surf on its beach than there would be on that of an inland lake half a mile across. On the opposite side, however, a heavy surf comes in, and a section of the beaches would look something like a series of steps. The highest beach is about 12 feet above high water mark. Between the two we have three distinct beaches. Of these the last is undoubtedly recent, the result of an unusually heavy storm. The next is comparatively recent, for it is not yet covered with grass and other vegetation, except *Mertensia, Honkenya,* and a few similar beach plants, which also cover a part of one below.

Thus we have nine successive beaches, gradually rising from the west to the east.

At least one other similar set of beaches is found on the island. These are in a curve of high lands, partially protected from the action of the surf. The beaches are much more numerous, and rather more irregular than in the preceding case.

Evidence to the same effect is found in a water-worn archway of rocks near the southern end of Amaknak. It is about 10 feet wide and 15 feet high, and passes through a wall of rock 50 feet high. It shows every evidence in its smoothed sides and graveled floor of being made by the action of waves, but its floor now stands 10 feet above high water mark, and is fairly out of reach of the highest surf.

Recent sea-urchins, shells, etc., are often found on the rocky hills 50 to 500 feet above the water level, but they have, in most cases, been undoubtedly brought there by birds. The writer has often seen ravens carrying them to such places. They raise the shells and similar objects 40 or 50 feet above the surface, and then drop them on rocks to break them open. The presence of such objects, except in strata, could hardly be taken as evidence of the elevation of the land.

Captain Hennig, one of the agents of the Alaska Commercial Company, who has lived in the Territory some years, informed the writer of the following facts: A harbor on Atka Island, one of the Middle Aleutians, in which not many years ago there was plenty of water, is now so completely shoaled that boats cannot enter. An island just south of the point of Alaska, formerly separated from the peninsula, is now connected with it by a neck of land four feet above high water. The writer has

heard in general terms, and from several sources, that many of
the passes between the Aleutian Islands, formerly safe, have now
shoaled so much as to have become dangerous. Dall, in his Re-
sources of Alaska, states that Isanotski, marked as a navigable
though dangerous pass between Unimak Island and Aliaska Peninsula by French surveyors, is now a *cul-de-sac.*

Dall also gives the following facts : On St. Michael's Island,
in Norton Sound, and on the neck between Norton and Kotzebue
Sounds, lie great winrows of drift wood, similar to those thrown
up to-day, but far beyond the reach of the water at its present
level. On St. Michael's Island also are some basaltic rocks, full
of amygdaloid cavities. The upper portion of the rocks is fully 15
feet above the level of high water, and some grass grows on it.
Yet in its cavities, *in situ*, can be found fragments of species of
barnacle, which must have lived there when the rock was covered daily by the tide.

Areas of local upheaval and depression occur in Southern
Alaska quite frequently, but they, of course, have but little bearing on the general question. A few years before our residence
there, a part of the site of the village of Illiuliuk, on Unalaska
Island, was lowered until covered by water, and a part of the
bay's bottom was brought to the surface. This occurred during
a severe earthquake. Dall quotes Captain Riedell to the effect
that a part of the south harbor of Unga Island, one of the Shu-
magin group, shoaled from 4 fathoms to 4 feet during an earthquake shock in May, 1868. It is well known that Bogosloff
Island, to the west of Unalaska, appeared above the surface with
much fire, smoke, tremblings of the earth and other disturbances
between May 1 and May 14, 1796. At this time, stones were
thrown to Umnak, a distance of about twenty-five miles. Eight
years after, when the island was visited, the sea was still hot
around it. Dall records the sinking of a low point in Chalmer's
Bay, Prince William's Sound. The stumps of the trees formerly
covering the point are now beneath the level of the lowest tides.
This is an isolated fact, and the phenomenon seems entirely local.

The northern shore of Alaska has the characteristics of a
land just rising from the sea. It is generally level and slopes

gradually toward the ocean. The latter is so shoal as to be dangerous for vessels for a long distance off the coast. From the coast inwards the land abounds in lagoons, inlets, lakes, all shallow. The evidence of this character is supported by the statement of Howorth (Nature, V., 163) that parts of the coast described by Beechey as cliffs, are now separated from the water by low flats.

As to other circumpolar lands, Reclus states that the southern end of Greenland is sinking, and then quotes Hayes to the effect that the northern end is rising along with Grinnell's Land. Hayes noticed sea beaches on these coasts that had been raised to the height of 100 feet. He also noticed that the rocky headland cliffs were polished by ice up to this height.

Going westward, Dall states that the eastern coast of Siberia is rising. Howorth quotes Wrangel and others to show that the northern coast of Siberia is rising. Drift wood is found 12 feet above the level of the sea, and large birch logs are scattered over some of the northern plains 3° north of any known Siberian forest. The coasts are low and flat, and a line of high ground, parallel with the sea coast, and representing a former beach, lies a few versts inland. Whales have deserted that part of the Arctic waters since the 18th century, probably owing to their shoaling. The Tundra or great Siberian plain is coated with fine sand like that on the coast now. Where Sanypcheff found comparatively deep water in 1787, are now found shoals and banks.

As to the Scandinavian Peninsula, the evidence is voluminous and positive. Not only, as Sir Charles Lyell has said, are Sweden and Norway rising, but the northern end is rising more rapidly than the southern. The evidence is given at considerable length in Lyell's "Principles of Geology."

Spitzbergen is also being elevated. Reclus gives the evidence that the islands of this group generally exhibit a series of beaches up to a height of 147 feet, with bones of whales and recent shells.

Alaska comes in, then, with her share of evidence for the theory, broached by Howorth, that the Arctic lands are rising. •

Another correspondent of Nature (V., 225), Mr. J. J. Murphy, suggests that the Antarctic lands are also rising. Still another correspondent, Geo. Hamilton, tries to show that the earth being an ellipsoid, a uniform shrinkage would result in a change of form and an apparent elevation in the circumpolar regions.

B.

LIST OF LAND AND FRESH-WATER SHELLS FOUND WITHIN A CIRCUIT OF FOUR MILES ABOUT ANN ARBOR, MICH.

CLASS, *Gasteropoda.*—ORDER, *Prosobranchiata.*

FAMILY, *Melaniidæ.*

Goniobasis Milesii Lea, (Huron River).
Goniobasis Levesens Mkc., (Huron River).

FAMILY, *Valvatidæ.*

Valvata tricarinata, Say, (Huron River.)

FAMILY, *Viviparidæ.*

Melantho integra, Say, (Still water, Huron River).

FAMILY, *Rissoidæ.*

Amnicola porata, Say, (Huron River).
Amnicola lustrica, Say, (Huron River).
Pomatiopsis Cincinnatensis, Anth., (River banks).

ORDER, *Pulmonata.*—FAMILY, *Helicidæ,*

Macrocyclis concava, Say, (School Girl's Glen).
Limax campestris, Binney, (Common under dead wood).
Helix albolabris, Say, (common).
Helix alolabris var. dentata, (occasional.)
Helix alternata, Say, (common).

Helix elevata, Say, (Dead specimens in recent deposits).
Helix exoleta, Binn., (Cascade Glen).
Helix fallax, Say, (common).
Helix hirsuta, Say, (common).
Helix labyrinthica, Say, (common).
Helix lineata, Say, (not abundant).
Helix monodon, Rack., (common).
Helix monodon, var. Leaii, (common).
Helix monodon, var. Fraterna, (common).
Helix multilineata, Say, (common).
Helix multilineata, var. albina, (uncommon).
Helix palliata, Say, (uncommon).
Helix perspectiva, Say, (south and west, uncommon).
Helix profunda, Say, (Cascade Glen).
Helix pulchella, Mul., (common).
Helix, solitaria, Say, (Dead). Live specimens found up the River by Dr. A. B. Lyons, Detroit.
Helix striatella, Anth., (common).
Helix thyroides Say, (common).
Helix tridentata, Say, (common).
Cionella subcylindrica, Leach, (common).
Pupa armifera, Say, (common).
Pupa contracta, Say, (common).
Pupa pentodon, Say, (South).
Pupa fallax, Say, river banks, (rare).
Vertigo milium, Gld., (common).
Vertigo ovata, Say, (common).
Succinea avara, Say, (common).
Succinea obliqua, Say, (uncommon).
Succinea ovalis, Gld. not Say, (uncommon).
Succinea Peoriensis, Wolf, (common).

FAMILY, *Arionidæ.*

Zonites arborea, Say, (common).
Zonites fuliginosa, Griff., (Cascade Glen).
Zonites indentata, Say, (common).
Zonites nitida, Mull., (common).

Zonites viridula, Mke., (common).
Zonites ligera, Say, (common).
Zonites minuscula, Binney, (South).
Zonites fulra, Drap., (common).
Tebennophorus Carolinensis, Bosc., (not abundant).

FAMILY, *Auriculidæ.*

Carychium exiguum, Say, (common).

FAMILY, *Limnæidæ.*

Limnæa columella, Say, (rare).
Limnæa humilis, Say, (common).
Limnæa stagnalis, Linn., (rare), lakes west and Huron River.
Limnæa palustris, Mull., (Swamp north-east).
Limnæa desidiosa, Say, (common).
Physa gyrina, Say, (common).
Physa gyrina, var. hildrethiana, (common).
Physa Sayii, Tappan, (small lakes west).
Physa heterostropha, Say, (common).
Bulinus hyphnorum, Linn., (North-east, common early in June).
Planorbis campanulatus, Say, (rare).
Planorbis trivolvis, Say, (common).
Planorbis bicarinatus, Say, (common).
Planorbis exacutus, Say, (common).
Planorbis albus, Mull., (rare).
Planorbis parvus, Say, (common).
Planorbis deflectus, Say, (not abundant).
Segmentina armigera, Say, (common).
Ancylus tardus, Say, (common).
Ancylus parallelus, Hald., (?) (uncommon).

CLASS, *Chonchifera.*—(SECTION, *Asiphonida.*)
FAMILY, *Unionidæ.*

Unio multiradiatus, Lea, (common).
Unio gibbosus, Barnes, (common).
Unio luteolus, Lam., (rare).
Unio verrucosus, Barnes, (common).

Unio pressus, Lea, (rare).
Unio novi-eboraci, Lea, (abundant).
Margaritana marginata, Say, (abundant).
Margaritana deltoidea, Lea, (not abundant).
Anodonta edentula, Say, (common).

(SECTION, *Siphonida.*)—FAMILY, *Corbiculadæ.*

Sphærium sulcatum, Lam., (common).
Sphærium occidentale, Prime, (Northeast swamp, common).
Sphærium partumeium, Say, (Huron River).
Sphærium striatinum, Lam., (Huron River, common).
Sphærium secure, Prime, (Northeast swamp, on the River).
Pisidium virginicum, Bourg., (River, common).
Pisidium variabile, Prime, (Huron River).
Pisidium compressum, Prime, (Huron River).
Pisidium abditum, Hald., (Huron River).

———

SUMMARY.

BRYANT WALKER,
CHAS. E. BEECHER.

UNIVERSITY OF MICHIGAN, ANN ARBOR, June 3, 1875.

C.

INDIAN MOUNDS IN GENESEE COUNTY.

The mound which is the subject of this paper, is in the town of Argentine, on the farm of L. C. Fletcher. It is a low mound twenty feet across, on the edge of a hill which overlooks a marsh. The descent to the marsh is gradual. The soil of the ridge is a coarse gravel resting on a substratum of clay and gravel, much harder than the superstratum.

The bottom of the mound rests on this stratum of clay and gravel, three feet below the surrounding level. The earth was piled above this level to form the mound, which had worn down to about three when discovered.

An oak tree, about one foot in diameter, had been growing on the center of the mound, but is now entirely gone. The mound was first opened in the center, when an entire cranium was found with its under jaw. This skull had double teeth in full set, and when closed on the under jaw left a space as if worn out by a pipe-stem. These bones were very much decayed, easily broken, and were the color of the soil.

Subsequently an examination of the mound was made, which resulted in the discovery of the remaining bones of the skeleton, a small urn, splinters of bone, some pieces of flint, including one perfect arrow-head, and three more craniums. The urn, flints, etc., were found under the first opening, evidently belonging to the first skull dug up. There seemed to have been but one entire body deposited, as the bones to the last three craniums were missing.

The urn is small, has a round bottom, will hold perhaps one-half pint. The pieces of bone were small, well preserved, quite smooth and tough ; none of them were over five inches in length.

Of the three craniums, one was entire; of another, only the frontal bone remained; the third had the frontal, parietal, and occipital bones. The right squamous suture of the last was crushed in. Within the skull was found a thin, sharp stone, about two inches long and one and a quarter inches wide. This evidently had entered edge uppermost, and narrow end first. These three skulls, like the first, had their faces to the east, were one behind the other, and about two feet apart.

Some of these articles are in the Museum of the University of Michigan.

One mile and a half south from this barrow, on the west shore of a lake, and the south bank of a river that makes out of the lake, are three mounds, all larger than the one first mentioned, and, perhaps, four or five rods apart. I have the word of two men who dug at different times, that these mounds contained pottery, flint implements, and one skeleton of immense size, such that the lower jaw fitted over the face of the man who dug it up.

Southwest of these is another, the largest of all. Report says this mound contained skeletons in a lying posture, and three rows deep, one layer above another.

Another class of remains found here is circular pits, formerly from three to four feet deep, and four feet in diameter. They were used for fire, as they are filled with charcoal, burnt sand, and a few traces of burnt bone.

LORENZO V. FLETCHER.

June 1st, 1875.

D.

THE AROMATIC GROUP IN THE CHEMISTRY OF PLANTS.

BY ALBERT B. PRESCOTT,

PROFESSOR OF ORGANIC CHEMISTRY IN THE UNIVERSITY OF MICHIGAN.

[Read before the Ann Arbor Scientific Association, July 3, 1875.]

The term Aromatic Group was first given to the benzoic series of compounds, as classified around the common nucleus benzoyl, by Leibig and Wöhler in 1833. Benzoyl, C_7H_5O, was the first compound radical recognized in complex vegetable products, and its discovery at once opened new ways of investigation in the field of organic chemistry. From time to time other chemical nuclei have been defined, in the construction of other groups of carbon compounds, until, now, it may be said, there are as many series of these bodies as most persons care to number among their scientific acquaintances. But there has been no greater activity or keener enthusiasm or richer reward, in all the labors of the forty years of organic chemistry, than in the work devoted to the aromatic group. The place which this group holds in organic chemistry is similar, in certain respects, to the position which organic chemistry itself occupies in general chemical science.

Among the valuable results of the labor devoted to this group we must accept, first, a clearer insight into the constitution of molecules, throwing better light upon the chemistry of all bodies. That phase of " the new chemistry" which finds expression in graphic formulæ—the theorizing as to the relations of atoms to each other within the molecule, with whatever of

4

well-grounded philosophy or of fallacious hypothesis appertains
—has, in no small share, been due to work which has been skep-
tically designated as that of "those German chemists running
crazy with what they call their aromatic group." It is not to be
supposed that the expenditure of this labor, or of any pioneer
labor in science, has been without waste. But the evidence of
its substantial success is before the chemical world in the long
list of well-defined aromatic bodies now as truly under the con-
trol of the chemist, in analysis and in synthesis, as are the metallic salts. And this evidence is not addressed to the chemical
world alone. The world of factories and ships, the world seek-
ing a sign as to the truth and use of all and any science, has received from the chemistry of the aromatic group a good number
of palpable demonstrations of the power of chemical knowledge.
There have been produced, under chemical direction, from the
waste of coal-gas manufacture alone, aromatic substances as follows : since 1856, anilin dyes, now sold at ten millions of dollars yearly, to color stuffs in the tints of the rainbow for every
household; since 1870, madder dye, amounting in 1873, to
1,000 tons, valued at over four millions of dollars ; and, this year,
the acid of wintergreen oil, promising to be the most useful of
the antiseptics, being applicable to foods and drinks,—beside a
considerable number of other products, in themselves of no
slight importance in commerce and the arts. Assuredly, the aromatic bodies have been found valuable material both in physical
science and in industrial economy.

As regards their significance in biological science, the question will arise : how far may an insight into the constitution of
molecules formed in plants help the chemist toward an understanding of the *formative steps* in plant chemistry? Before we
can reach this final question in our subject, we must consider,
first, what the aromatic group is and where in the vegetable
kingdom it extends, and, next, by what steps the molecules of this
group are formed outside of living bodies, under conditions arranged by the chemist.

The aromatic group, when this term was first adopted, consisted of benzoic acid, bitter almond oil, amygdalin, cinnamic

acid, cinnamon oil, cuminic acid and cummin oil, with a con-siderable number of artificial derivatives from these bodies,—nearly all having penetrating aromatic odors. The possession of aromatic odors, however, is not especially characteristic of the very great number of bodies now classified in this group. The present definition of the aromatic group, is, in briefest terms, *the series of bodies built upon benzene.*

Benzene (also termed benzole), containing $\frac{12}{13}$ carbon and $\frac{6}{13}$ hydrogen, may be formulated as C_6H_6: being in vapor 39 times heavier than its volume of hydrogen, its molecular weight is 78 and it must be formulated as C_6H_6. Carbon, here as almost everywhere, is a tetrad, and the six tetrad atoms of car-bon have twenty-four bonds of chemical union. As only six of these are occupied by the six monad atoms of hydrogen, eighteen bonds must (it is believed) connect carbon with car-bon,—forming nine lines (or movements?[1]) of union between carbon atoms. As it is found in certain compounds that each one of the six carbon atoms behaves alike, it appears that each must hold the same relations to its fellows, and they must be dis-posed in a ring, as first proposed by Kekule, in 1865. The nine lines of union between the six atoms of carbon (if not connecting alternate or opposite atoms, *i. e.*, if disposed in the ring) require the alternate unions to be made by double lines. Then each atom of carbon in the ring has one bond of union free to be held by atoms (or semi-molecules) outside of the carbon ring: so that as regards other elements each atom of carbon is a monad.

By substitutions, for one or more of the six hydrogen atoms, of other (monad) atoms or radicals, the formulæ of the various aromatic compounds are obtained: the graphic symbol being always a hexagon.

(1) KEKULE, *Ann. Chem. u. Pharm.*, clxii, 77. LADENBURG, *Deut. Chem. Ges. Ber.*, v., 322. *Watts' Dictionary of Chem.*, 2nd Supplement, 132.

The substitution of methyl, CH_3, for one, two, three, etc., of the atoms of H (around the hexagon)—and also for atoms of H in CH_3—constructs *the hydrocarbons* of this group :

$$C_nH_{2n-6}$$ *Number of possible Isomers.*

1. C_6H_6 Benzene.
2. $C_6H_5(CH_3)=C_7H_8$ Toluene.
3. $C_6H_4(CH_3)_2=C_8H_{10}$ Xylene.   Three($1, 2 : 1, 3 : 1, 4$).
4. $C_6H_3(CH_3)_3=C_9H_{12}$ Cumene.   Three($1, 2, 3 : 1, 2,$
                                                         $[4 : 1, 3, 5)$.
5. $C_6H_2(CH_3)_4=C_{10}H_{14}$ Cymene.   Three($1, 2, 3, 4 : 1,$
                                                         $[2, 4, 5 : 1, 3, 4. 5)$.
6. $C_6H(CH_3)_5=C_{11}H_{16}$ Amylbenzene. One($1, 2. 3, 4, 5$).
7. $C_6 (CH_3)_6=C_{12}H_{18}$ Amylmethylbenzene.
etc.

          4
    5   3
    6   2
          1

The hydrocarbon having two of the original hydrogen atoms displaced (no. 3) may evidently have contiguous or alternate or opposite atoms displaced, thus presenting three different kinds of molecules,—and the number of variations mathematically possible is given above as a theoretical number of possible isomers. The full number of possible isomers, on this theory, has been actually produced in most instances though not in all, and perhaps has not been exceeded in any instance. Thus, there are known three compounds having the ultimate composition and molecular weight of xylene, differing in certain properties: orthoxylene (having parts 1 and 2 of the hexagon occupied by methyl), isoxylene (having 1 and 3) and metaxylene (having 1 and 4 occupied by methyl). This correspondence between fact and theory in the number of isomers strengthens the evidence that the hexagonal figure represents the *actual relations* of the atoms in the molecule. But it must be borne in mind that, while figures are placed upon paper to represent certain ascertained *relations* of atoms to each other, nothing has been ascertained as to the *places* of atoms in molecules.

Other relations than those of place may be represented geometrically, with advantage.

The aromatic hydrocarbons, together, are known as the benzoles—the chief liquid distillates from coal tar—coal tar naphtha. Among the distillates from coal tar are two other hydrocarbons, which are solid, and are of unusual importance in industry, namely : naphthalene and anthracene. These are not homologous with benzene, as members of the same series ($C_n H_{2n-6}$), but belong each in another series related progressively to the benzene series, with mathematical harmony, as follows :

Benzene —— $C_6 H_6$ or $C_n H_{2n-6}$ —— Single Hexagon——

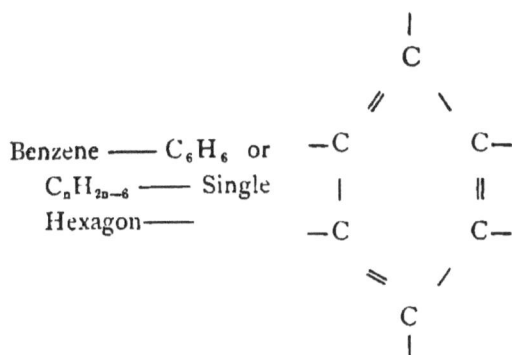

Naphthalene——$C_{10} H_8$ or $C_n H_{2n-2 \times 6}$ —— Double Hexagon—

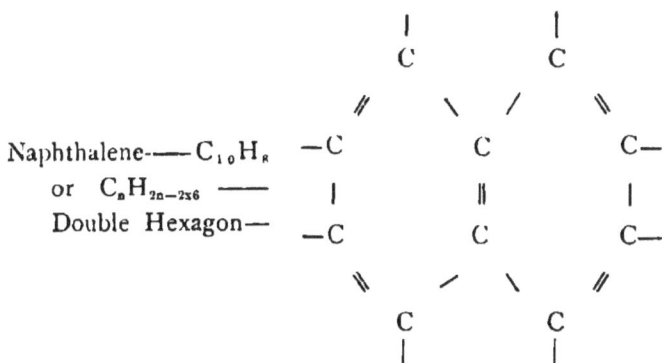

$$
\begin{array}{ccc}
\mid & & \mid \\
\text{C} & & \text{C} \\
/\!/ \quad \diagdown \ / & & \diagdown \\
-\text{C} \quad \quad \text{C} & & \text{C}- \\
\mid \quad\quad\quad \parallel & & \mid \\
-\text{C} \quad\quad\quad \text{C} & & \text{C} \\
\diagdown\!\diagdown \ / \quad \diagdown \ /\!/ & & \diagdown \\
\text{C} \quad\quad \text{C} & & \text{C}- \\
\mid \quad\quad\quad \mid & & \parallel \\
\quad -\text{C} & & \text{C}- \\
& \diagdown\!\diagdown \ / & \\
& \text{C} & \\
& \mid &
\end{array}
$$

Anthracene—--$C_{14}H_{10}$ or C_nH_{2n-3x6}—--Triple Hexagon—--

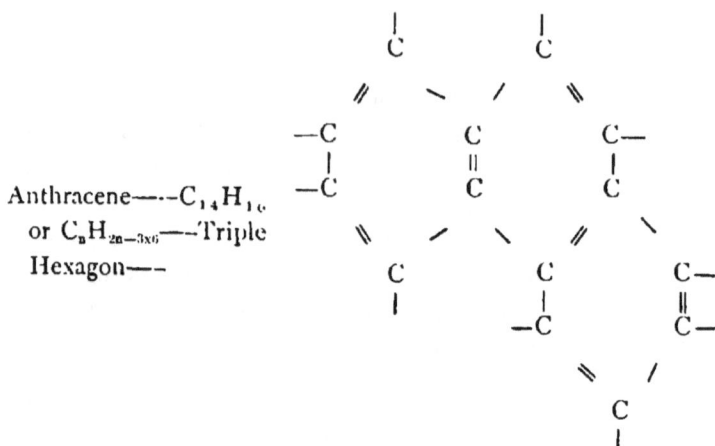

From these hydrocarbons (*i. e.*, from the hexagon, single, double or triple) all the aromatic bodies are extended. *Displacing* (one or more atoms of)

H by OH,   phenols are formed (as in carbolic acid).

H by $C\begin{Bmatrix} O \\ OH \end{Bmatrix}$.   acids "   "   (as benzoic acid).

$\left.\begin{matrix} H \\ H \end{matrix}\right\}$ by $\begin{Bmatrix} OH \\ C\begin{Bmatrix} O \\ OH \end{Bmatrix} \end{Bmatrix}$, acids "   "   (as salicylic acid).

H by $N\begin{Bmatrix} H \\ H \end{Bmatrix}$, amines   "   "   (as anilin).

$\left.\begin{matrix} H \\ H \end{matrix}\right\}$ by $\left.\begin{matrix} O \\ O \end{matrix}\right\}$, quinones   "   "   (as anthraquinone).

The aromatic hydrocarbons have been looked upon as bodies of too simple chemical construction to exist in plants, and this is certainly true of the lighter portion of them. The first three members of the benzene series are not found in the vegetable kingdom, and the fourth, cumene, or trimethylbenzene, has been reported found only in Roman cummin oil. But the fifth member of the series, cymene, or tetramethylbenzene—a body having the molecular weight 134 and hence in vapor 67 times heavier than its bulk of hydrogen—a liquid closely approaching both in composition and in properties to turpentine oil—is, in its various isomers, distributed among plants to an extent not fully understood. It has generally been put down as more especially an educt of three plants, cuminium cyminium (cummin) and cicuta virosa (water hemlock), of the umbelliferæ, and thymus vulgaris[2], of the mint family ; also sometimes of a fourth, ammi copticum. But a large number of the volatile oils of plants contain hydrocarbons of the composition $C_{10}H_{14}$, more and more of which are found by chemical treatment to yield aromatic products and almost certainly to be built upon the benzene nucleus and to fulfil the character of "cymenes." Eucalyptus globulus, the Australian fever tree, which has received much attention of late years, contains a cymene as well as a terpene.

Now there appears to be only a short step in composition between cymene, $C_{10}H_{14}$, and *oil of turpentine*, $C_{10}H_{16}$, which is found in the coniferous trees, but this short step suffices to throw oil of turpentine out of the homologous series of aromatic hydrocarbons. Now isomeric with turpentine oil proper are " the terpenes" generally, including the essential oils of apricot, bergamot, birch, camomile, caraway, cloves, cubeb, elemi, hop, juniper, lavender, lemon, orange, parsely, pepper, savin, spike, tolu, thyme, and an indefinite number not named or not brought to notice. In fact, a large proportion, probably a majority, of essential oils contain terpenes, generally with other essential oil constituents. To present some approximate indication of the extent of distribution of the volatile oils altogether, a count has been made of the number of plants reported to contain volatile oils in the tabular summary of plant constituents given as the Second Part of Wittstein's Chemischen Analyse von Pflanzen und Pflanzentheilen (1867). This summary includes 576 plants in 114 natural orders. Among these, essential oils are reported in 156 or 27 per cent. of the plants, and in 45 or 39 per cent. of the families.[3]

---

(2) C. R. A. WRIGHT, *Jour. Chem. Soc.*, 1873, 686.

(3) Of the natural orders, the Labiatæ had the largest number of plants containing volatile oils (13 p. c. of all); the Umbelliferæ the next largest number; the Symantherlæ next, and the Myristiceæ next (6 p. c. of all).

Oil of turpentine and its numerous isomers have mostly
been placed, unclassified, among " the vegetable substances little
known," but there is a beginning that promises to draw them
into the aromatic group and assign them graphical formulæ of
the the hexagonal type. In 1872, A. Oppenheim reported to
the German Chemical Society an investigation at the Berlin
Laboratory[4] on the production of a cymene from oil of turpentine by abstraction of hydrogen (cymene dibromide being heated
with anilin). As a result of his research the investigator gives
this graphical formula for turpentine oil : Here one of the carbon
atoms of the ring has been loosened
and displaced by the triad CH, giving
the adjacent carbon atom only one line
of connection in the ring and two outside bonds, so that four carbon atoms
carry five of hydrogen. The remaining two points of the hexagon have
taken methyl (CH_3) instead of hydrogen and one of these methyl molecules
has methyl again substituted for two of
its hydrogen atoms. Later, Oppen-

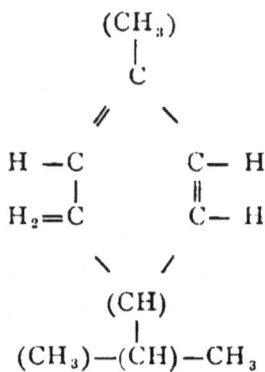

$$(CH_3)$$
$$|$$
$$C$$
$$/\!/ \qquad \backslash$$
$$H - C \qquad\qquad C - H$$
$$| \qquad\qquad |\!|$$
$$H_2 = C \qquad\qquad C - H$$
$$\backslash \qquad /$$
$$(CH)$$
$$|$$
$$(CH_3) - (CH) - CH_3$$

heim reports the formation of cymene from the terpene of cummin oil[5] and offers other confirmations. Kekule has since
traced relations between cymene and camphor (the immediate
oxidate of a terpene), $C_{10}H_{16}O$, from which he presents a
graphical formula. Kekule confirms the relationship between
cymene and the terpenes, using iodine instead of bromine, and
accepts Oppenheim's conclusions.[6]

Graebe has dissented from the view that turpentine oil is of
the aromatic type, although he finds a relation between the terpene in wormseed oil and cymene.[7] The conclusions of Oppenheim are mostly confirmed and extended by the labors of C. R. A.

(4) *Berichte der deutschen chemischen Gesellschaft*, V., 91.

(5) *Berichte d. d. chem. Gesell.*, v, 628; vi., 915.

(6) *Bericht. d. deut. chem. Gesell.*, vi., 437: *Jour. Chem. Soc.*, 1873, 889.

(7) *Deut. chem. Ges. Ber.*, v., 677.

Wright with oils of lemon, orange and nutmeg, reported to the London Chemical Society, in 1873.[8] It seems, then, strongly probable, if not yet fully established, that the terpenes, extending as they do widely through the vegetable kingdom, are all built upon the benzene nucleus, being, however, somewhat complex extensions of that nucleus.

From this standpoint, too recent to be secure, no small share of the most simple carbon compounds which plants contain—those destitute of oxygen and nitrogen—belong to the aromatic group. It has long been well known that plants contain *oxidized* aromatic compounds very closely related to hydrocarbons, so that the latter are easily *obtained from plant constituents.* Benzene itself took its name from the abundant educt benzoic acid, from which Mitscherlich first obtained it in 1834, by heating with lime or by stronger heat with iron. It is obtained by dry distillation or by more gentle treatment from the most of the aromatic compounds, and by harsher methods from bodies *not* built upon it; having been discovered by Faraday in 1825 among the vapors distilled at a high temperature from fats. Toluene was obtained by Deville, who first examined it and named it, by distilling Tolu balsam; and it is obtained with especial ease by the moderate action of heat upon a large number of resins—this being one of the many indications that most resins contain the aromatic nucleus. Xylene is also easily obtained from resins. The reverse of these transformations, the constant production of resins from terpenes by atmospheric oxidation in balsams and turpentines, will require mention further on. And the manufacture of aromatic hydrocarbons, on the large scale, from coal, will be noticed with the methods of production from inorganic sources.

After the hydrocarbons, we next inquire as to the distribution of the *oxidized products* of the aromatic group: phenols, acids, aldehydes, etc.

---

(8) *Jour. Chem. Soc.*, xi., 686-701.

The *phenols* are formed by substitution of OH for H attached directly to the C of the ring. The first phenol, known as purest grade of carbolic acid, is obtained by gentle decomposition of many plant constituents. Cymophenol, $C_{10}H_{13}OH$, is found in the oil of thyme, from the Gymnospermæ. Of the diatomic phenols, pyrocatechin, $C_6H_4(OH)_2$, is readily obtained from tannins by distillation and exists ready formed in Ampelopsis Hederacea. Creosote, obtained by destructive distillation of many bodies, contains two homologous phenols, diatomic and triatomic, each bearing methyl :

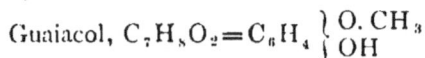

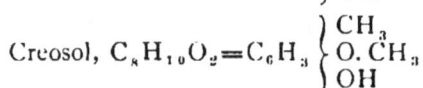

—OH

$$\text{Guaiacol, } C_7H_8O_2 = C_6H_4 \begin{cases} O.CH_3 \\ OH \end{cases}$$

$$\text{Creosol, } C_8H_{10}O_2 = C_6H_3 \begin{cases} CH_3 \\ O.CH_3 \\ OH \end{cases}$$

Pyrogallol (pyrogallic acid, largely used as a deoxidizing agent by photographers) is a triatomic phenol, $C_6H_3(OH)_3$, and is readily formed from tannins and from gallic acid; while its isomer, phloroglucin, is obtained from resins and from glucosides by heating with potassa.

The orcins, isomers of $C_7H_8O_2$, substitutions of two molecules of hydroxyl and one of methyl for three atoms of hydrogen in benzene, are found ready formed in lichens. The dyes archyl, cudbear and persio contain orcins,—as also litmus, from Leconora Tartarca. Aloes, treated with potassa, yields orcins.

And now, in 1872, Schiff ascribes to *gallic acid*, $C_7H_6O_5$, a rational formula consisting of the introduction of one carboxyl and three hydroxyl molecules in place of four of the hydrogen atoms of benzene.[9] Farther, he presents and

$$C_6H_2 \begin{cases} C \begin{cases} O \\ OH \end{cases} \\ OH \\ OH \\ OH \end{cases}$$

maintains a rational formula for fermentable *tannic acids*, the natural source of gallic acid.[10] Having at last synthesized tan-

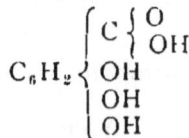

(9) *Ann. Chem. Phar.*, clxiii, 299; *Jour. Chem. Soc.*, 1872, 829.

(10) *Jour. Chem. Soc.*, 1872, 245, 1019, 1008,; 1874, 267.

nic acid from gallic acid by working in accordance with his theory, it cannot be ignored. Schiff also traces a very interesting correspondence between the astringent acids and other aromatic compounds, in the bright colors which great numbers of them give with ferric salts; the most familiar instance being that of ink, and instances being familiar to analysts, in the identification of phenol, benzoic, salicylic and cinnamic acids, etc. The wide distribution of the tannins gives great significance to this insight into their construction. Of the 576 plants in Wittstein's summary, 35, or 6 per cent., are given as containing tannic acids; these plants being found in 17 per cent. of the natural orders.

The quinones claim attention in this inquiry, for though they are not found in plants they are next steps to plant constituents. They hold oxygen atoms united to each other, as before stated. Ordinary quinone, $C_6H_4(O_2)''$, is easily obtained from the quinic acid of the cinchonas and from the allied caffeic acid of coffee,—being in each case a means of recognition in analysis. Resins, by heat, yield umbelliferone, an isomer of quinone.

The quinone of anthracene (the triple-hexagon nucleus) is called anthraquinone, $C_{14}H_8(O_2)''$. From this, by displacing H_2 with $(OH)_2$, is now manufactured dioxyanthraquinone, the *alizarin of madder*, as hereafter to be described. Isomeric with alizarin is the chrysophanic acid found in rhubarb, senna, and the wall lichen (parmelia parietina). Chrysammic acid, which is formed from the aloin of aloes in the common nitric acid test, also from the chrysophanic acid of rhubarb in the same way, has the composition of tetranitra-dioxy-anthraquinone, $C_{14}H_2(NO_2)'_4(OH)'_2(O_2)''$.

Taking next the *acids* derived from the benzene nucleus, with their aldehydes and alcohols, we have first benzoic acid and bitter almond oil, the well-known subjects of Liebig and Wöhler's first advance into the aromatic group. Always classed with these are salicylic, cinnamic, cuminic and anisic acids, aldehydes and alcohols; the aldehydes being essential oils of plants and the acids being the products of the natural oxidation of the aldehydes. In the following list, *each dash prefixed indicates the displacement of one atom of hydrogen from benzene*, C_6H_6, to the residue of which the additions are made. Thus, benzoic acid is $C_6H_5 C \begin{cases} O \\ OH, \end{cases}$ etc., (the H of OH in acids being replaced by metals, forming salts).

	Acid.	*Aldehyde.*	*Alcohol.*
Benzoic	$-C \begin{cases} O \\ OH \end{cases}$	$-C \begin{cases} O \\ H \end{cases}$	$-C \begin{cases} H \\ H \\ OH \end{cases}$
Salicylic $\begin{cases} \\ \\ \\ \end{cases}$	$-OH$ $-C \begin{cases} O \\ OH \end{cases}$	$-OH$ $-C \begin{cases} O \\ H \end{cases}$	$-OH$ $-C \begin{cases} H \\ H \\ OH \end{cases}$
Cinnamic (Phenyl-acrylic) $\Big\} \Big\{$	$\begin{cases} - \\ (C_3H_3)''' \end{cases} \Big\} \begin{matrix} O- \\ O \quad O \end{matrix} \Big\}$	$(C_3H_3)'''$	$\underset{(C_3H_3)'''}{-\!-} \Big\} \begin{matrix} H \\ H \\ O \end{matrix}$
Cuminic (Phenyl-propylic) $\Big\} \Big\{$	$-(C_3H_7)'$ $-C \begin{cases} O \\ OH \end{cases}$	$-(C_3H_7)'$ $-C \begin{cases} O \\ H \end{cases}$	$-C_3H_7)'$ $-C \begin{cases} H \\ H \\ OH \end{cases}$
Anisic $\begin{cases} \\ \\ \end{cases}$	$-O(CH_3)$ $-C \begin{cases} O \\ OH \end{cases}$	$-O(CH_3)$ $-C \begin{cases} O \\ H \end{cases}$	$-O(CH_3)$ $-C \begin{cases} H \\ H \\ OH \end{cases}$

Benzoic acid is accompanied in the balsams,([13]) and often complemented by cinnamic acid. It is formed from its aldehyde, bitter almond oil, by exposure of the latter to air. It is now manufactured from the naphthalin of coal tar (the double molecule benzene before described), by the process mentioned farther on. Benzoic aldehyde is hardly an educt; being a product, along

---

(13) Styrax benzoin, Myrospermum toluiferum, Myrospermum peruiferum, Eunonymous Europeus.

with hydrocyanic acid and sugar, in the natural fermentation of amygdalin which is found in many plants of the almond family.

Salicylic acid does not exist uncombined in plants, that the writer is aware, but as methyl salicylate it makes the principal portion of "wintergreen oil," from Gaultheria Procumbens and Betula Lenta (sweet birch) and occurs in great purity and abundance in Andromeda Leschenaultii([14]). This acid, the new antiseptic, is now being manufactured on the large scale from carbolic acid, as presently to be described. It is not poisonous ; 1¼ gramme doses being taken without apparent ill effect. It prevents most fermentive and putrefactive changes; including those, like the sinapous and amygdalous, which are not dependent upon an organized ferment, as well as the alcoholic and lactic. One-tenth per cent. prevents grape juice from fermenting, and 0.04 per cent. delays the souring of milk 36 hours later than when the milk is not so treated([15]). It arrests putrefactive changes, as well as fermentive. Unlike carbolic acid, its antiseptic power is destroyed by alkalies. The methyl salicyate has been has been to some extent manufactured for use instead of natural wintergreen oil. *Salicylic aldehyde* is known as the oil of spiræa, found in "meadow sweet" and "hardhack." It is readily obtained by fermentation of the glucoside *salicin*, the bitter substance of the willow and poplar and found with its product in meadow sweet. Populin and Helicin both readily yield salicin.

Cinnamic acid—phenyl-acrylic—is found in the balsams ; and appears when its aldehyde cinnamon oil is exposed to the air. The balsams also contain,—in styrax, cinnamic alcohol, cynnyl cinnamate ($C_9H_9C_9H_7O_2$), and cinnamene, C_8H_x. Cinnamate of benzyl ($C_7H_7C_9H_7O_2$) forms a large part of Peru balsam and a small part of Tolu balsam.

Cuminic aldehyde is found, with cymene, in cummin oil (from C. Cyminium).

(14) BROUGHTON: *Phar. Jour.*, Oct. 7, 1871.

(15) NEUBAUR: KOLBE: MULLER: *Jour. Chem. Soc.*, 1875, 159, 160.

The *Anisic series* is closely related to the oil of anise (Pim-
pinella Anisum and Illicium A.).

The benzene nucleus has not been traced in many of *the
alkaloids.* Atropia, however, yields an isomer of cinnamic acid,
and it is conjectural that other alkaloids of the Solanaceæ contain the benzene nucleus.

In the Summary of Wittstein, before mentioned, $\frac{26}{114}$ or 22
per cent. of the natural orders contain aromatic bodies—terpenes
and resins not being included as such.

After this slight survey of the constitution and natural distribution of the aromatic bodies, we will next consider what has
been accomplished in their *artificial synthesis.*

Benzene itself, it will be observed, is a polymer of acetylene, C_2H_2. And by heat, in a bent tube over mercury,
acetylene (13 times denser than hydrogen) is transformed into
benzene (39 times denser than hydrogen). Acetylene is formed,
from the elements carbon and hydrogen, in the electric arc,
with a strong battery (one of 40 or 50 Bunsen's elements giving
100 c.c. of the gas per minute[16]). Also from marsh gas, or
from carbon disulphide with carbon monoxide, by electric dis-
charge.

Toluene is formed from benzene, (1) by action of methyl
iodide and sodium (Fittig and Tollens), (2) by marsh gas when
both it and the benzene are nascent (Berthelot)[17]. In the last
reaction, Xylene and Cumene are also produced.

In these and other ways all *the hydrocarbons* expressed with
the single hexagon may be synthesized from benzene. Naphthalene, of the double hexagon, is formed on passing toluene through
a white-hot tube (Berthelot)[18].

It is familiar to every one that these hydrocarbons, with the
phenols and many other aromatic compounds, are formed every
day in every town *from coal, by distillation* at ten or twelve hun-

(16) BERTHELOT: *Compt. Rend.,* IV, 640; *Watts' Dict.,* 1st Sup., 30, 31.

(17) $C_6H_6 + CH_4 = C_7H_8 + H_2$.

(18) $4\,C_7H_8 = C_{10}H_8 + 3\,C_6H_6 + 3\,H_2$.

dred degrees Fahrenheit, and this, it is submitted, is organic synthesis. Coal is a very simple if not completely mineralized mixture of carbon with bituminous hydrocarbon, and it seems a misnomer to style as "destructive distillation" the formation of the aromatic group from such material. It is *formative* distillation : prolific beyond parallel in the action of heat upon elements.

It may be safely stated in general terms that the formation of the aromatic compounds, from the elements, outside of living bodies is assured. And there are now at least four aromatic substances manufactured on a large scale from coal-tar ; one being anilin and its homologues, and the other three being vegetable educts, benzoic acid, alizarin, and salicylic acid.

The manufacture of *benzoic acid* from napthalin is carried on through two steps, namely :

1. Oxidation, by hot nitric acid, to phthalic and oxalic acids.[19]

2. Removal of the elements of carbonic anhydride from phthalic acid, by heating with lime in a close vessel.[20]

| Naphthalin. | Phthalic acid. | Benzoic acid. |

Successive substitutions.*

In the manufacture of alizarin from anthracene, there is :

1. Oxidation to anthraquinone.

(19) $C_{10}H_x + 8O = C_xH_6O_4 + H_2C_2O_4$.

(20) $C_xH_6O_4 = C_7H_6O_2 + CO_2$.

* One side of the double ring is broken up; its four points of CH being oxidized to four semi-molecules of carboxyl, two of which enter into the phthalic acid, the other two uniting as oxalic acid.

2. Farther oxidation to dioxyanthraquinone.[21]

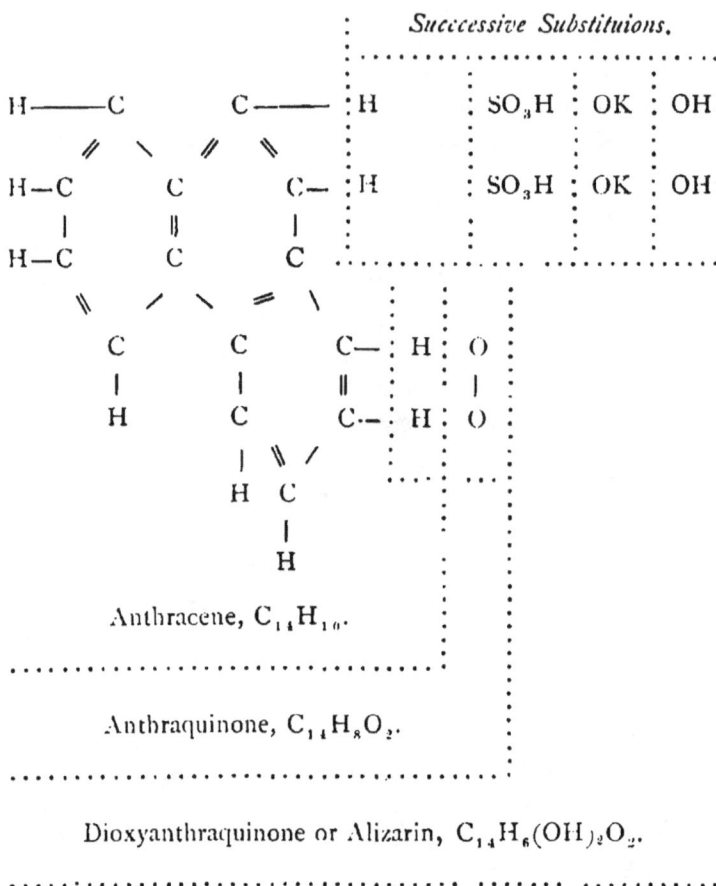

			Successive Substitutions.			

$$
\begin{array}{cccccccc}
\text{H} & \longrightarrow & \text{C} & \text{C} \longrightarrow & \text{H} & \text{SO}_3\text{H} & \text{OK} & \text{OH}\\
 & & & & & & & \\
\text{H}-\text{C} & & \text{C} & \text{C}- & \text{H} & \text{SO}_3\text{H} & \text{OK} & \text{OH}\\
\text{H}-\text{C} & & \text{C} & \text{C} & & & & \\
\end{array}
$$

$$
\begin{array}{ccc}
\text{C} & \text{C} & \text{C}-\ \text{H}\ \text{O}\\
\text{H} & \text{C} & \text{C}-\ \text{H}\ \text{O}\\
 & \text{H}\ \text{C} & \\
 & \text{H} &
\end{array}
$$

Anthracene, $C_{14}H_{10}$.

Anthraquinone, $C_{14}H_8O_2$.

Dioxyanthraquinone or Alizarin, $C_{14}H_6(OH)_2O_2$.

The history of this triumph may be indicated as follows :[22]

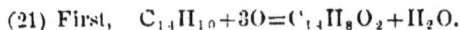

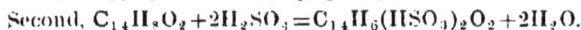

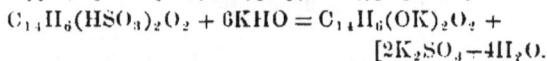

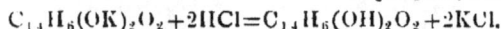

---

(21) First, $C_{14}H_{10}+3O=C_{14}H_8O_2+H_2O$.

Second, $C_{14}H_8O_2+2H_2SO_3=C_{14}H_6(HSO_3)_2O_2+2H_2O$.

$C_{14}H_6(HSO_3)_2O_2+6KHO=C_{14}H_6(OK)_2O_2+[2K_2SO_4-4H_2O$.

$C_{14}H_6(OK)_2O_2+2HCl=C_{14}H_6(OH)_2O_2+2KCl$.

(22) Artificial production of Alizarin: Roscoe: *Chem. News*, xxi., 185; Amer. Reprint, p. 332 (1870).

Purpurin, the less important color compound of madder, having the ultimate composition of trioxyanthraquinone, is this year reported to be synthesized from alizarin.[23]

A ton of alizarin is saved from about 2,000 tons of coal or 160 tons of coal tar. By this saving, large agricultural districts in Holland and Alsace and in Asia Minor, employed hitherto in producing color, can be devoted to the production of food.

The production of *salicylic acid* from phenol, through action of carbonic anhydride, is simply a direct synthesis: a molecule of carboxyl (CO_2H) being substituted for one of the hydrogen atoms of phenol. That is, the elements of a molecule of carbonic acid gas, CO_2, are added to the elements of a molecule of phenol, C_6H_6O, to form a molecule of salicylic acid, $C_7H_6O_3$.

$$
\begin{array}{c}
H \\
| \\
C \\
\diagup\!\!\diagdown \\
H-C \qquad C-\quad H \\
| \qquad\quad \| \\
H-C \qquad C- \quad H \quad C\begin{Bmatrix} O \\ OH \end{Bmatrix} \\
\diagdown\quad\diagup \\
C \\
| \\
OH
\end{array}
$$

Phenol, C_6H_6O.

Salicylic acid, $C_7H_6O_3$.

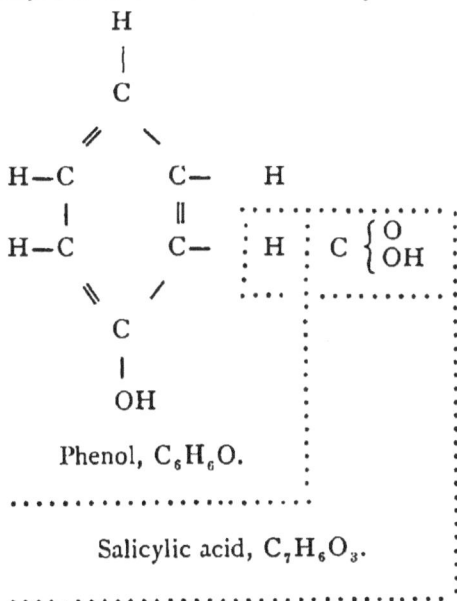

(23) LALANDE: *Compt. Rend;* lxxlx, 669; in *Jour. Chem. Soc.*, 1875, 69.
ROSENTIEHL: *Compt. Rend.*, lxxix, 761; in *Jour. Chem. Soc.*, 1874, 373.

This change was sometime since effected by Prof. Kolbe, through the action of carbonic acid gas on phenol in presence of sodium. Last year he gave preliminary notice of his present method, in which soda at elevated temperatures takes the place of metallic sodium, and early in the present year the manufacture of salicylic acid from carbolic acid and sal soda by use of carbonic acid gas commenced at Leipsic ([24]).

It is singular that the reverse of this change, the manufacture of pure carbolic acid from the salicylic acid of wintergreen oil, was reported on by Broughton, the quinologist of the British Government in India, in 1871, as possibly a remunerative enterprise, at least "in case of war" or other occasion of increase in the English price. The oil was obtained from the Andromeda Leschenaultii, which grows in great abundance on the Neilgherry Hills ([25]).

These few instances of artificial synthesis in the aromatic group have acquired prominence on account of their relations to wealth and industry; but instances of almost equal scientific importance are thickly spread among the reports of every month.

The natural production of aromatic bodies does not wholly elude chemical investigation. Some of the chemical changes in the aromatic constituents of *the balsams* are striking illustrations of the well known characteristics of these bodies. Resins are produced from the terpene and cymene oils by atmospheric oxidation; and benzoic and cinnamic acids are produced from the aldehyde. alcohol, and ether of the cinnamic series, by oxidation; and without doubt these oxidations occur within coniferous trees, as well as more generally after exudation from the bark, and in the same way in the bottle on the shelf where light falls. As changes of oxidation, these are not representative of the vegetable kingdom; nevertheless they are in the direction of greater complexity of chemical structure.

---

(24) KOLBE: *J. pr. Ch.* [2] viii, 41; In *Jour. Chem. Soc.,* 1874, 373.

(25) *Phar. Jour.* and *Trans.* Oct. 7, 1871.

The production of resins from terpene and cymene oils, now that these are seen to be aromatic hydrocarbons, explains the ease with which other aromatic bodies (benzene, toluene, phenol, etc.) are obtained from resins,—for it can now be more than surmised that in this round of changes the benzene ring is never broken.

In the summary of Wittstein, of the 114 natural orders, 28 are reported to contain resins; of these 28 orders, 16 contain essential oils with the resins and 12 do not. On the other hand, there are 45 orders containing volatile oils, 29 of them not being reported as containing resins. Of the 26 orders given as furnishing aromatic bodies other than resins and oils, one-half have resins.

To enter fully into an inquiry as to *the chemical history of the aromatic bodies in plants* would be to overstep the limits of this paper. Indeed, it may be thought that such an inquiry would overstep the present limits of science. Let us consider what preparation we have and what foundation we have for entering upon such an inquiry.

In the first place, we have some measure of acquaintance with the structure (or to be more modest, the chemical character) of aromatic compounds. We know something as to what a given aromatic substance can be formed from, and what can be formed from it, and the conditions needed in both cases. This knowledge is demonstrated by the large number of *syntheses* which chemical science has effected in the aromatic group. But we must be cautious about assuming that substances producible in the laboratory in a certain way must needs be formed in the plant in the same way. We must recollect that we have already often observed that a given chemical production may be effected in different ways. There are, well known, at least three different ways of bringing gallic acid out of gallotannic acid : fermentation by the natural ferment, "fermentation" by boiling, and oxidation. Bruise a bitter almond kernel with water, and by reason of the emulsin present, bitter almond oil and prussic acid arise in vapor and the solution becomes sweet with glucose.

Again, boil the almond pulp with dilute sulphuric acid, and the bitter almond oil and prussic acid and glucose appear with another product, formic acid. Vinegar may be rapidly formed from alcohol, in the air, (1) when at ordinary temperature there is the contact of a certain species of living cells; (2) when, without cells, there is platinum black present; (3) when, without cells or porous body, the oxygen is nascent; the change being in each instance through aldehyde by the same equation. Now, if the styrax benzoin contained naphthalin, it would not certainly follow that the benzoic acid of the plant was formed from this naphthalin, through phthalic acid, because such is the case in the factory.

In the second place, as foundation for a study of the chemical history of aromatic bodies in plants, we have but a very limited knowledge of the constituents of plants in general. The *analytical work* in organic chemistry is behind the synthetical work. The proximate analysis of plants needs to be made as thorough as possible: no constituent can be assumed to be unimportant. As an illustration, it was stated above that in a certain summary of plant constituents, of 45 orders reported to contain volatile oils, 29 were not reported to contain resins. Considering the known methods of analysis and the ordinary purposes of analysis, the question arises, how many of the plants analyzed in these 29 orders do nevertheless contain resins? And, taking a given plant known to contain both resin and volatile oil, what results might not come from a series of careful quantitative analyses of the plant in different stages of its growth and of different parts of the plant? Before the natural formation of carbon compounds can be traced, and before generalizations as to nature's chemical methods can be attained, an enormous amount of work has to be done.

In this discussion, it is of course taken for granted that *the molecules of matter are formed and conserved by chemism;* as truly in the plant as in the rock. Chemism may be, as has been held, due to an attractive force; or it may be due to harmonies and coördinations of atomic motion, rotatory or oscillatory; molecules may be plane or solid forms, in gaseous or solid state; and it may be that we have in no case attained any correct conception as to what causes molecular combination; but, none the less, the effects we know, and their consummate order we know under the name of chemical law. For myself, this is quite enough. I have a profound sense that the cause of chemical action is beyond the comprehension of man and is near to the hand of God. I do not see that the chemist can assume any more responsibility for the construction of molecules in the test-tube than can the biologist for the growth of cells under the microscope. If we could see and measure the molecules, we should doubtless be no nearer the comprehension of their formation than the biologist is to the formation of cells under his inspection. But whatever be the scope of the human mind in chemical science, it is a science that embraces all that we can know of the composition of matter. In living tissue, the elements (which in mixture would be but dust and gases) are *combined* into certain kinds of matter, and this combination (with transformation) fulfills the definition of chemical union.

In limitation, it hardly need be remarked that when chemical action has formed the molecule it can do no more: it cannot make *the cell*, or any other structure or coherent mass formed of molecules, any more than the cell-making action can make the molecule. Now, as cohesion and as heat and other actions impel or retard or modify chemical action, it seems almost certain to be true that the action of cell organization must impel or retard or modify chemical action. We see that red-hot charcoal will, with oxygen, form carbonic anhydride, while cold charcoal will not: the heat is indispensable, but we do not conceive the carbonic anhydride to be a calorific compound. It is a chemical compound, whatever non-chemical actions are essential to its formation. And, if it should ever be settled that certain substances can only be formed in living cells, it is submitted that these substances must none the less be accepted and studied as chemical products.

E.

DR. COCKER'S ABSTRACT OF PAPER ON LIFE.

[Not furnished to the Committee in time for publication.]

F.

EXTRACTS FROM, AND A SUMMARY OF, A PAPER "ON LIFE."

BY E. C. SEAMAN.

[Read before the Association December 4th, 1875.]

Dr. Beale inquires, what is life? Very few intelligent persons will mistake the phenomena of life—or the difference between inanimate and living matter; and yet it is not easy to give a definition of life, that will be satisfactory to all the schools of physiology and biology. In my opinion there is in the material world, a *vital element, separate and distinct from all the other elements of nature—which is an organizing principle, and constitutes the essence of life,* and causes the internal action of living organisms.

The definitions of life generally given by authors, include phenomena only,—which are only the effects of the action of life, operating through the machinery of a living organism. My definition of life is, that it is the power of a vital organizing element, which forms and maintains living organisms—which produces internal action—including in animals, digestion and absorption, assimilation and secretion, nutrition and reproduction.

The whole action of the animal economy, including digestion and secretion, assimilation and nutrition, and also the circulation of the blood, is carried on antagonistically to chemical action, as well as to gravitation.

The vital element must float in the gastric juice and in the blood, in the chyle and in the nervous fluid, and pervade the flesh and all the solids of the living organism. When particles cease to be animated with the vital element, they are soon secreted, and eventually excreted from the system; and if any obstruction be offered to their secretion or excretion, they are turned into other chanels, and either inflammation, or an abnormal formation is the result.

The phenomena of life indicate that the vital element is very subtle imponderable matter, in some respects similar to caloric and electricity—except that it has no more mobility than oxygen, or any other gas. Like caloric and electricity, it may be supposed to float in the atmosphere and in the ocean, and to pervade the crust of the earth; to operate upon, unite, and to form into germs and organisms, atoms of matter in the atmosphere, in the earth, and in the waters upon the earth. It is the organizing principle and the cause of the formation of the germs, and of the development of plants and animals de novo, and without parentage, in water, at the bottom of the ocean, and upon newly formed islands.

I regard as self-evident truths,—1st, That force is an inherent property of substance, and cannot exist independent of substance.

2d, That no combination of elements can produce a force not inherent in any of the elements themselves.

3d, That the vital force, not being the subject of chemical analysis, must be an inherent property of an element of matter unknown to the chemist.

Prof. Tyndall and the advocates of the dynamical school of scientists, deny the existence of a vital element, and maintain that the vital force is identical with what they call molecular force. In my view, it is a pure assumption, and imaginary, to suppose that there is any such thing as a molecular force distinct from, or other than, the chemical and physical forces of the elements of which the molecules are composed.

Dr. John Hunter said, " The living principle of the blood is the *materia vitæ diffusa,* of which every part of an animal has its portion. It is diffused through the whole solids and fluids, making a necessary part of them."

In Dr. Prout's Bridgwater Treatise (p 493), he says : " The stomach must have the power of organizing and vitalizing the different alimentary substances. It is impossible to imagine that the agency of the stomach can be chemical. This agency is vital, and its nature is unknown."

Dr. Pritchard says : " This vital principle assumes the character of a plastic or formative power. It presides over and sets in action the different processes by which growth and organization are effected ; and gives form and modification to the component parts of the animal and vegetable body, and contributes, by a preserving influence to the maintenance of its existence, for a definite portion of time."

The vital element acts antagonistically to the forces of gravity—it must therefore be imponderable. The vital force carries the sap of trees and plants, and the blood of animals, upwards— contrary to the forces of gravity inherent in the ponderable elements of matter ; and hence living organisms must have, in their composition, some imponderable element, unknown to the chemist.

Prof. Nicholson, of the University of Toronto, says, in his Biology, " It appears in the highest degree probable, that every *vital action* has in it something which is not merely physical and chemical, but which is conditioned by *an unknown force— higher in its nature and distinct in kind,* as compared with all other forces. The presence of this *vital force* may be recognized even in the simplest phenomena of nutrition ; and no attempt has hitherto been made to explain the *phenomena of reproduction,* by the working of any known physical or chemical force."

Dr. Beale of London says, " Facts and observations on things living, support the idea of *vitality,* and are not favorable to any mechanical or chemical hypothesis of life yet proposed."

The material organism of animals is composed of elements of matter known to the chemist ; but the vitalists hold that in addition to the chemical elements, there is a vital element, which pervades the whole organism, and works the machinery of life ; that the chemical elements constitute only the frame work of the living organism, which cannot operate itself without a motive power ; and that the vital force inherent in a vital element, constitutes the motive power, which works the machinery of the animal economy—the mind constituting the governing power which works the muscles, and produces voluntary action.

Vital action produces cells and tubes for the circulation of the fluids, which form and support animal and vegetable organisms. It produces the peculiar organic forms, of which cells and tubes and fibrous matter form prominent parts. On the contrary, chemical action never produces either cells, tubes, or fibers, nor organisms of any kind, having the machinery necessary for internal action, nutrition, and reproduction.

Chemical compounds, whether liquids, solids, or gasses, have a composition nearly uniform in all their parts, and never have anything like cells, tubes or fibrous organs. Their solids are generally composed of crystals, like the crystals of common salt. Chemical compounds also increase from the outside—by the addition of other crystals to the mass. ' On the contrary, living organisms always increase from the inside—the food being received internally, and digested, absorbed, secreted, and conveyed in a liquid form, through tubes and cells to the proper places of deposite, to nourish the system. The difference in formation and action indicates that there must be elements in one not contained in the other.

The *vital theory* was presented in a more complete, clear and distinct form, by Dr. Hunter of England, the latter part of the eighteenth century, than ever before. My reading indicates that the *vital theory*, in some of its forms, was generally accepted by Physicians and Physiologists, until after the publication of Baron Liebig's Oganic Chemistry ; and that it has been received ever since, and is still received as the *true theory of life*, by Physiologists of the first rank. Dr. Beale does not stand alone in affirming its soundness.

ON SPONTANEOUS GENERATION.

EXTRACTS FROM, AND SUMMARY OF, A PAPER READ BEFORE THE ASSOCIA-
TION,

BY E. C. SEAMAN.

Prof. Dunster, in his paper on the History of Spontaneous Generation, read before the Association March 4th, gave an account of many changes of opinion among learned men and scientists upon the subject ; and stated many facts and thories, and referred to numerous experiments, made by persons investigating such questions—both in Europe and America ; all of which was and is valuable as well as interesting, to inquirers into such questions. But I must be allowed to express my dissent from the deductions and conclusions of the Professor. He informs us that the French Academy, some years since, referred the question of spontaneous generation, and numerous experiments of M. Pouchet in support of it, and of M. Pasteur against it, to a committee, which, after a long and careful examination, reported, that the experiments apparently proving the truth of the theory were not made with due care and proper precautions, and therefore were not reliable ; and that those of M. Pasteur tending to disprove it, were reliable ; and Prof. Dunster insists that the committee virtually settled the matter, as a scientific question ; and that it should now be considered as settled by science and scientists, against the theory.

During a period of more than 2,000 years, from the time of Aristotle to sometime in the 18th century, the doctrine of spontaneous generation was generally received in Europe as a truth, beyond all dispute. During the last hundred and fifty years various ous theories have been conceived and presented, to account for the origin of the lower orders of animal and vegetable organisms ; and so far from the question being settled by scientific investigations and scientists, there never was any previous period in the world's history, when reading and learned men and scientists were so much divided in opinion, and embraced so many conflicting theories upon the subject of life and its origin, as they do to-day.

In most of the experiments made and reported by Pouchet, Wyman, Bastian, and others, animal life was developed from water and vegetable infusions, which had been heated to the boiling point, hermetically sealed in glass jars, and exposed for some time in a warm atmosphere. Pasteur boiled vegetable infusions from 4 to 6 hours, sealed and exposed them in a similar manner, and no animal organisms were developed.

Heat and water are the great solvents of nature. Steam and hot water not only destroy life, but soon dissolve all animal fibres and organs, except bones. Because water and vegetable infusions will not develop animalcules after having been boiled from 4 to 6 hours, it does not follow, that animal life is never originated without the agency of parents. Such boiling not only destroys the germs and incipient organisms existing in the water and in the vegetable matter, but it drives out the vital element itself, and destroys the very essence of the vegetable matter, evaporates the nutritious portions of it, and unfits it for food, for the development of other organisms. Nature never boils the materials out of which she developes living organisms.

Living organisms may originate :

1st, By a special creation of God.

2d, By the spontaneous action of the elements and forces of nature, without the agency of parents, *which is called spontaneous generation;* or

3d, By the spontaneous action of the elements and forces of nature, *acting through the agency of parents, and their ova, sperm, spores, germs or seeds.*

Parentage does not supersede, but only co-operates with, the elements and forces of nature, in propagating and producing living organisms; and when they have come into existence, the agencey of parents soon ceases, and their future growth and maintenance depends wholly upon the elements and forces of nature. These are great and important truths, as well as facts, which every reader and investigetor of this complicated subject should bear in mind.

The production de novo and without parentage, of an animal or vegetable organism, can arise only under favorable circumstances ; but I want no clearer, nor more complete evidence than my own observation, and the facts and experiments reported by Pouchet, Bastian, Prof. Wyman, and numerous other persons, and the results of their experiments, to establish to my satisfaction, the doctrine of spontaneous generation. The errors of Pasteur and the French Academy, arise from a misinterpretation of the phenomena of nature, from misinterpreting the experiments and their results. The doctrine of spontaneous generation rests upon the *vital element, and the vital force inherent in it.*

The theory of Pasteur, known as Panspermism, is, that the atmosphhere, and bodies of water also, are pervaded with animalcules and their eggs, germs or spores, from which are developed all the animalcules found in vegetable infusions.

Since the assumed settlement of the question by the experiments of Pasteur and the decision of the French Academy, Dr. Bastian of London, after years of study and experiment, with the aid and light of the experiments of Pasteur and Pouchet, Wyman, and many other scientific inquirers, published in 1872, two volumes upon *The Beginnings of Life*, in which he discusses at length, the subject of life and its origin, the various modes of reproduction, and spontaneous generation. He states and comments upon many theories, and hundreds of experiments of himself and others, points out the errors and fallacies of Pasteur and the French Academy, and affirms that cases of spontaneous generation do frequently occur.

Dr. Bastian shows most clearly, 1st, That a temperature of 212° F. for even one minute, is sufficient to destroy the vitality of any of the lower organisms ; and I may add, that it requires but a short time to destroy their texture. 2d, That very few animalcules or their ova, germs or spores, can be found in the atmosphere, even in summer ; and I apprehend that none can be found in winter, in high latitudes ; 3d, That multiplication by the ordinary process of reproduction, will not adequately account for the thousands of ciliated infusoria, often met with in the course of a few days, in many organic infusions.

These objections, supported by numerous experiments, should be regarded as sufficient to dispose of the Panspermic theory of Pasteur, as unsound and fallacious.

No animalcules, flies, insects, or their ova, can exist in the atmosphere in winter, in high latitudes, without being destroyed. Hence we have no flies and no insects in cold weather. The experiments reported by Prof. Wyman were made in winter ; and hence the swarms of animalcules that were developed could not have come from atmospheric germs ; but must have been generated spontaneously, from the water and vegetable matter which he used.   Prof. Wyman says :

" After the flasks were prepared, they were suspended from the walls of a sitting room, near the ceiling, where they were exposed to a temperature of between 70 deg. and 80 deg. F., throughout the day, and nearly the same during the night."

After stating the details of each experiment, the materials used, and its results—showing that in different flasks vibrios, bacteriums, monads, and other species of animalcules were produced, often in large numbers, the Professor says :

" We have here a series of thirty-three experiments, prepared in different ways, in which solutions of organic matter, *some of them previously filtered have been boiled* at the ordinary pressure of the atmosphere, for a length of time, varying from 15 minutes to two hours, and exposed to air purified by heat."

" In many instances, a solution like that in sealed flasks, and boiled for the same length of time, was exposed to the ordinary air of the room, in an open flask.   Although the same forms were found in the two, they appeared much more rapidly in the open than in the closed vessel."

" The result of the experiments here described is, that *the boiled solution of organic matter* made use of, *exposed only to air which had passed through tubes heated to redness*, or enclosed with air in hermetically sealed vessels, and *exposed to boiling water, became the seat of infusorial life.*"

The questions of life and its origin are problems which science can never settle with certainty ; for like every thing relating to force, as well as to intellect, they are to man profound mysteries—belonging to a large extent, to the unknown and the unknowable. Force, and the essence of life and of intellect, not being visible to the eye—being beyond the power of the chemist and the microscopist to discover, are not the subjects of positive science. Each of them is the subject of inference only, and of uncertain human reasoning. Hence the variety of opinions upon such questions. But being matters of inference, from the phenomena of nature and the action which we witness in the universe, as well as from the consciousness which every person has of the action and cognitions of his own mind, we have some means of inquiring into the nature of such mysterious causes and processes.

To talk about *molecules and protoplasm, molecular forces and dynamical forces*, throws no light upon the subject ; but tends to involve it in mysticism.

By experiments and careful observation of natural phenomena—of physical, vital, and intellectual action and their results, together with long study and inquiry into the causes of such action and results, all our physical and metaphysical sciences have been built up—including physiology and zoölogy, botany and medicine, as well as physics and chemistry. Much has been learned in relation to each and all of them—however imperfect our knowledge may still be. Many theories more or less false, but the most of them partially true, have been conceived, generally received for a time, and then superseded by others. And thus man has groped his way through darkness, to the present state of human knowledge—often led astray by false assumptions, and false theories—arising from erroneous inter-pretations of the action of natural causes and of the elements of nature. His inquiry into subjects which he can never master, and can never attain to positive and certain knowledge, has been profitable, and productive of valuable results. It has enabled him to attain some ideas and knowledge of the mysteries of nature and of God. But to pretend that such imperfect and obscure knowledge is positive science, would be an absurdity.

The forces of nature being the inherent properties of the elements of matter, it seems impossible that any one element should contain properties and forces inconsistent with each other. Though the forces of nature are various, and many of them conflicting and antagonistic, all antagonisms must come from different elements, and not from antagonistic forces in the same elements. Though all action and effects must have causes, to suppose that acts widely different in their nature, can be produced by the same causes, and those only, or that combinations different in their nature can be composed of the same elements, and those only, is to suppose what is palpably inconsistent, and therefore impossible. To suppose that crystals and living organisms can be produced by the same causes, and out of the same elements, and those only, involves a gross inconsistency—the one being nearly solid in their form and texture, and the other permeated with canals and tubes, in which fluids circulate, to nourish and maintain the organism. Life can never result from any combination of elements of inanimate matter.

The action of the animal economy and of mind cannot be rationally accounted for, without supposing, that there are in existence elements, which neither the chemist nor the surgeon can detect, and which the microscopist cannot see ; and hence, to account for the wonderful powers of the human mind, we must infer, from the action and phenomena of mind, that there is in the brain of man, *an intelligent spirit, distinct from the chemical elements of which the material organization of the brain is composed;* and to account for the origin, growth and maintenance of living beings, and the action of the animal economy, we must infer, from their manifestations and effects, that they have a vital organizing element, endowed with properties and forces very different from those of inanimate matter, of which earths and rocks, salts and crystals are composed.

Such inferences are rational, because they are necessary to account for the phenomena. Natural phenomena constitute the evidences, from which the inferences are drawn, and upon which they are based. They are not mere assumptions, as the material-ists alledge, but rational interpretations of natural phenomena—logical conclusions drawn from competent evidence. Every man's consciousness testifies to himself, of the existence of an intelligent spirit within him, which acts in a manner very differ-ent from the ponderable elements of the body. Such evidence is all which the nature of the case admits of, and to reject it, because it is not equal to a geometrical demonstration, would be absurd. If such evidence be rejected, neither intellectual nor physical science can have any foundations to rest upon, or evi-dence to support them. If such evidence be rejected as unreli-able and untrue, there can be no truth in any thing, that is called science, beyond mere abstract mathematics.

Caloric is the great stimulant of nature, without which neither chemical nor vital action can take place. Chemical action originates spontaneously and de novo, under proper cir-cumstances, and when the temperature is right, to stimulate it; but will not operate very actively, when the temperature is much below summer heat.

Chemical action, fermentation, putrefaction, and decompo-sition, are all cases of the spontaneous action of the elements and forces of nature, when the heat is sufficient, and all the cir-cumstances favorable. The vital forces, co-operating with the physical forces, carry on the machinery and processes of the ani-mal economy; and why cannot the same forces, acting under favorable circumstance, form spontaneously the proper elements into living germs and organisms, and thus originate plants and the simpler forms of animals? It is as easy to conceive how such elements and forces can originate and form de novo plants and animals, as it to conceive how they can be propagated and developed from eggs, sperm, seeds, germs or mere buds. No matter how animals and plants may be propagated or produced, the processes are mysterious:—beyond the reach of scientific analysis—and beyond the reach of any thing like positive and certain science. The processes of the propagation and development, growth and maintenance for years, of animals and trees, are no less mysterious than their spontaneous generation.

The physico-chemical forces, sometimes called dynamical forces, produce spontaneously gases and water, crystals and salts, earths and ores, rocks and stones ; but they never produce either animal or vegetable organisms, without the *co-operation of the vital force*, which is an inherent property of a peculiar imponderable element of matter, unknown to the chemist. That unknown element, called *the vital element*, constitutes of itself the life of the blood of animals, the seed in the earth, spoken of in 1st Genesis, 11th and 12th verses—the life and active principle of the eggs, sperm, germs, seeds and buds, which develop into animals and plants.

11. "And God said, let the earth bring forth grass, the herb yielding seed, and the fruit tree yielding fruit after its kind, *whose seed is in itself, upon the earth;* and it was so.

12. "And the earth brought forth grass," etc.

The question arises, what was the seed spoken of in Genesis; and from whence come the animalcules and the germs of life, which the microscope reveals in the air, and in still water in summer ? Some affirm that they all spring from living parents ; and that all plants, grasses and weeds spring from seeds grown upon the stalks of parent plants, of like character. Can that be so ?

Grain and seeds kept dry will not freeze in any temperature ; but if saturated with water, they will freeze. Freezing and thawing not only destroys the vitality of animal and vegetable organisms, but soon destroys their texture, so that they decay when the hot weather comes. How can the swarms of mosquitoes of Greenland, Lapland, and other high latitudes be accounted for, unless they are produced spontaneously, by heat, moisture, and a vital force, acting upon the decayed vegetable and animal matter of previous years ? To suppose that frozen eggs will hatch and produce living animals, is to make a suppo-

6

sition contrary to reason—as well as to all human experience and observation.

Seeds in or on the ground either grow or rot, very soon after the earth becomes sufficiently warm and moist to promote vegetation. Seeds, planted in the fall or early in the spring, do not lie dormant until July or August, and then germinate and grow. Gardens in cities are generally kept so free from weeds that none are allowed to go to seed; and yet the following year, if the ground be rich and the season be wet and warm, thousands of weeds spring up—more in the latter part than in the fore part of the season; and it is impossible to account for them, unless they originate and grow spontaneously, and without seed. How quickly fire-weeds spring up in the woods where a log-heap or a pile of brush has been recently burned, and all seeds killed by the fire. Weeds which grow from seeds produced the previous year, usually spring up in May or early in June; but we know that in a warm and wet season, there are from five to ten times as many weeds springing up in gardens and in corn and grain fields and pastures, in the months of July, August and September, as during the months of May and June. Large quantities of summer grass and weeds grow in corn fields after the last cultivation of the corn in July. The fact that the most of the weeds that grow each year in cold and temperate climates, spring up after the first of July, constitutes unanswerable evidence that those coming up late in the season do not come from seeds of parent plants, but are spontaneous productions of the earth.

Mushrooms usually spring up in the latter part of the summer, upon heaps of manure which have been fermenting for weeks in succession; and upon droppings of manure upon old pastures. It is very evident, that the most of them do not come from seed of parent plants grown the previous year; for, if such were the case, they would come earlier in the season. I can see no reason to doubt, that they are the *spontaneous productions of a living principle,* acting upon the manure out of which they grow. And so it is, with the natural grasses and herbs, shrubs and trees of every country—with the mosses which grow upon trees, barren rocks, and upon the decaying wooden roofs of old buildings, as well as upon the buckets of wells—and also with the sea-weed growing at the bottom of the ocean.

Many facts, which constitute clear and satisfactory evidence of spontaneous vegetation, are presented by Prof. A. Winchell, late of the University of Michigan, in his work entitled "Sketches of Creation." On page 250, he says: "Nothing is a more common observation than to see plants making their appearance in situations where the same species was previously unknown, or for a long time unknown, and under circumstances such, that the supposition of a recent distribution of seeds is quite precluded."

Again he says: "Earth thrown out of cellars and wells is generally known to send up a ready crop of weeds, and not unfrequently, of species previously unknown in that spot. In all these cases (many being cited), after allowing for all known possibilities of the distribution of seeds by winds, birds, and waters, it seems probable that germs must have previously existed in the soil"

The Panspermists attribute the production of animalcules in vegetable infusions, put up and sealed in glass flasks, to germs and spores in the atmosphere, drawn into the flasks before they were sealed. The seeds of plants and trees, herbs and grasses do not grow and mature in the ground, where new plants originate. Do the germs from which they grow originate in the ground, as the Panspermists maintain that the germs of animalcules originate in the atmosphere? If such germs of plants, grasses, etc., originate in the ground, as they must, what can be the process, other than that of spontaneous generation? To account for the origin of plants and grasses in mysterious cases, the Panspermists must invent new fallacies, and present to the public new dogmas. Their old fallacies and dogmas will not answer the purpose. They will not bear the light of reason and common sense.

The germs of plants, not formed in seeds of previous plants, but originating in the ground, could not have had a parental origin ; they must have been generated in the ground, as animalcules are generated in the atmosphere, and in water of the proper temperature ; and it is impossible to conceive how they could be thus originated, without parentage, unless by spontaneous generation, or by a special creation of God. Do animalcules propagate in the atmosphere or in water ? Is it possible to conceive how animalcules, originating either in the atmosphere, in water, or in vegetable infusions, and particularly in winter, in cold climates, could have a parental origin ? My belief is, that they are produced spontaneously, by the elements and forces of nature.

Can any one conceive how herbs and grasses, trees and shrubs of various kinds, originated on islands of comparatively recent origin, and on mountains many thousand feet above the level of the sea—(that must have been thrown up by internal upheavals of the earth,) unless there have not only been innumerable special creations of seeds or original plants and trees, in millions of places, but also that such special creations took place at thousands of different periods in the world's history ; or 2d, that the elements and forces of nature can, and have, originated such organisms, and produced them spontaneously? If you suppose they all originated from special creations of God, the difficulty is only partially removed—the mystery still remains, how they germinate and grow—how they are preserved from the antagonistic influences of the chemical and physical forces,—and how they reproduce their kind, in diverse modes, through the agency of seeds, roots, slips and buds. Is it possible to form a rational conception of the causes of the various natural processes of living organisms, unless mother earth, and the elements and forces of nature, have power to originate and develop them de novo, as well as to preserve them, and cause them to propagate their kind?

Pope, in his Essay on Man, gives a rational interpretation to the 11th and 12th verses of 1st Genesis, when he says :

> " See through this air, this ocean, and this earth,
> All matter quick, and bursting into birth.
> Above, how high, progressive life may go!
> Around, how wide, how deep extend below!
> Vast chain of being! which from God began,
> Nature's etherial, human, angel, man,
> Beast, bird, fish, insect! what no eye can see,
> No glass can reach; from infinite to thee;
> From thee to nothing."

G.

FLORA OF ANN ARBOR AND VICINITY.

In accordance with the directions of the Association, the Committee on the Flora of this vicinity have made out a report. The report includes the Phænogams and three orders of Vascular Cryptogams.

The Committee were limited to a circle having a radius of four miles, Ann Arbor being the center. Within these limits we have represented one hundred and one (101) orders, three hundred and seventy-eight (378) genera, and eight hundred and forty-eight (848) species.

Dr. Gray, in his Botany of the Northern States, gives one hundred and thirty orders ; as already stated, we have one hundred and one represented, leaving only twenty-nine of which we have no representatives in this small tract of country. This shows the richness in variety, at least, of this vicinity.

We have appended lists of exterminated, rare and local. and introduced plants. Of exterminated plants there are fourteen (14) species. This number includes Nyssa multiflora or Pepperidge tree, the wood of which is very hard, and was formerly used for beetles and wedges ; also Dirca palustris, which was used by the Indians for thongs.

Of rare and local plants there are thirty-two (32) species, this number including Viola rostrata, which is called one of the rare violets. With us it is one of the most common. In this list of rare and local plants, there is one, viz. : Aplectrum hyamale, commonly called putty root, and also known as Adam and Eve, which will soon be added to the number of exterminated

plants, if the gentlemen from the University continue their dep-
redations. The following incident was related by the late Miss
Clark, and we repeat it as given by her. A number of the students wishing specimens of this plant for examination, applied to
Miss Clark to learn the locality. Upon being informed they
straightway proceeded to the place, and dug up all that was to be
found. About this time they were reminded of the nearness of
the dinner hour by the pangs of hunger which they experienced.
So, being without other food, they determined, for once, to let
the cause of science take care of itself and ate up all of the
specimens they had obtained. Miss Clark used to say, "that
when the faculty turned the boys out to grass, they should devise
some means by which the extermination of entire species might
be prevented."

By introduced plants, we mean plants which have made
their appearance within the last twelve or fifteen years. Of these
there are sixteen (16) species. In identifying one of these,
Thlaspi arvense, we have had some difficulty. Dr. A. B. Lyons
found it in fruit in 1867, Prof. Harrington found it in 1868, and
it was again found in fruit in 1869; but not until 1870 was it
found in flower. By this time the new edition of Dr. Gray's
Botany had been published, and in this we found a description
of the plant.

Among the introduced plants we have some which will give
trouble in the future if not soon exterminated. We will mention
two. 1st, Circium arvense or Canada thistle. This is spreading
by the roots. Our city authorities prevent its spreading from
seed by keeping the tops cut down. The other is Cenchrus tribuloides or bur grass. It made its appearence on the R. R. bank
near the City Mills in 1872, only a few plants at first. Now it is
spread over an acre or more.

MISS E. C. ALMENDINGER.

RANUNCULACEÆ.

Clematis Virginiana, L.
Anemone cylindrica, Gray.
Anemone Virginiana, L.
Anemone Pennsylvanica, L.

Anemone nemorasa, L.

Hepatica triloba, Chaix.

Hepatica acutiloba, D. C.

Thalictrum anemonoides, Michx.

Thalictrum dioicum, L.

Thalictrum purpurascens, L.

Thalictrum Cornuti, L.

Ranunculus divaricatus, Schrank, Huron River, Ann Arbor.

Ranunculus aquatitis, L. var trichophollus Chaix, Prof. M. W. Harrington, Huron River, Ann Arbor.

Ranunculus multifidus, Pursh.

Ranunculus, multifidus, Pursh, var terrestris, common in ditches at Tamarask Swamp.

Ranunculus abortivus, L.

Ranunculus sceleratus, L.

Ranunculus recurvatus, Poir.

Ranunculus Pennsylvanicus, L.

Ranunculus fascicularis, Muhl.

Ranunculus reprens, L.

Ranunculus bulbosus, L., found only in 1872.

Ranunculus acris, L., University campus, 1866.

Caltha palustris, L.

Captis trifolia, Salisb., wood South of city, Prof. M. W. Harrington.

Aquilegia Canadensis, L.

Hydrastis Canadensis, L. Rich Woods, not common.

Actæa spicata, L. var rubra, Michx.

Actæa alba, Bigel.

MAGNOLIACEÆ.

Liriodendron Tulipifera, L., Geddesburg.

ANONACEÆ

Asimina triloba, Dunal., one mile South of Ann Arbor.

MENISPERMACEÆ.

Menispermum Canadense, L.

BERBERIDACEÆ.

Caulophyllum thalictroides, Michx., Local.

Jeffersonia diphylla, Pers, one locality two miles up the river.

Podophyllum peltatum, L.

NYMPHÆACEÆ.

Brasenia peltata, Pursh., in the lakes West of Ann Arbor.

Nymphæa adorata, Ait.

Nuphar advena, Ait, in the lakes West of Ann Arbor.

SARRACENIACEÆ.

Sarracenia purpurea, L., Peat-bogs.

PAPAVERACEÆ.

Argemone Mexicana, L., Miss C. Watson.

Sanguinaria Canadensis, L.

FUMARIACEÆ.

Dicentra Cucullaria, D C., Found but once, Prof. M. W. Harrington.

CRUCIFERÆ.

Nasturtium afficinale, R. Br.

Nasturtium palustre, D C.

Nasturtium Armoracia, Fries.

Dentaria diphylla, L.

Dentaria laciniata, Muhl.

Cardamine rhomboidea, D C.

Cardamine rhomboidea, var purpurea, Torr.

Cardamine pratensis, L., Tamarack swamp and near Bunker's Dam.

Arabis hirsuta, Scap., North side of Huron River beyond first R. R. bridge West, 1861.

Arabis Canadensis, L.

Arabis Drommandii, Gray, Geol. Sum., 1860 and Prof. M. W. Harrington once.

Barbarea vulgaris, R. Br.

S isymbrium afficinale, Scop.

Brassica Sinapistrum, Boissier, Geol. Surv., 1860.
Brassica Alba, Geol. Surv., 1860.
Brassica nigra, Geol. Surv., 1860.
Camelina satina, Crantz, Road side Ann Arbor.
Capsella Bursa-pastoris, Mœnch.
Thlaspi arvense, L. under the Papaws.
Lepidurm Virginicum, L., Prof. M. W. Harrington.
Lepidurm intermedium, Gray, not common.

VIOLACEÆ.

Solea concolor, Ging, one locality two miles up the river.
Viola blanda, Willd.
Viola cucullata, Ait.
Viola cucullata, var palmata, West of Ann Arbor, 1861 ; not seen since.
Viola sagittata, Ait, damp ground East of cemetery ; on Campus ; now exterminated.
Viola pedata, L., Geol. Survey, 1860.
Viola canina, L., var sylvestris, Regel.
Viola rostrata, Pursh.
Viola striata, Ait, Under the Papaws.
Viola Canadensis, L., Under the Papaws.
Viola pubescens, Ait.
Viola var eriocarpa, Nutt.

CISTACEÆ.

Helianthemum Canadense, Michx.
Lechea major, Michx.

DROSERACEÆ.

Drosera rotundifolia, L., Peat-bogs around the lakes West of Ann Arbor.
Drosera longifolia, L., Peat-bogs around the lakes West of Ann Arbor.

HYPERICACEÆ.

Hypericum pyramidatum, Ait, Bank of the Huron River, near the second R. R. bridge east, 1866.
Hypericum prolificum, L.

Hypericum ellipticum, Hook, Geol. Surv., 1860.
Hypericum perforatum, L.
Hypericum corymbosum, Muhl.
Hypericum mutilum, L.
Hypericum Canadense, L, Geol. Survey, 1860.
Elodes Virginica, Nutt, Near the lakes West of Ann Arbor.

CARYOPHYLLACEÆ.

Sapanaria afficinalis, L.
Silene antirrhina, L.
Lychnis Githago, Lam, Wheat fields.
Arenaria serphyllifolia, L., State street, Ann Arbor.
Stellaria media, Smith.
Stellaria longifolia, Muhl.
Cerastium vulgatum, L.
Cerastium viscosum, L.

PORTULACACEÆ.

Portulaca obleracea, L.
Claytonia Virginica, L.

MALVACEÆ.

Malva rotundifolia, L.
Malva sylvestris, L.
Malva moschata, L.
Abutilon Avicennæ, Gaertn.
Hibiscus Trianum, L.　Geol. Survey, 1860.

TILIACEÆ.
Tilia Americana, L.

LINACEÆ.

Linum Virginianum, L.　Road side, Ann Arbor, Dr. A. B.
Lyons.

GERANIACEÆ.

Geranium maculatum, L.
Erodium cicutarium, L. Her.　May, 1871, Prof. M. W.
Harrington.
Impatiens fulva, Nutt.
Oxalis stricta, L.

RUTACEÆ.

Zanthoxylum Americanum, Mill.
Ptelea trifoliata, L.   Along railroad.

ANACARDIACEÆ.

Rhus typhina, L.
Rhus glabra, L.
Rhus venenata, D. C.
Rhus Toxicodendram, L.
Rhus Aromatica, L.

VITACÆ.

Vitis aestivalis, Michx.
Vitis cordifolia, Michx.   Geol. Survey, 1860.
Ampelopsis quinquefolia, Michx.

RHAMNACEÆ.

Rhamnus alnifolins, L. Her.
Ceanothus Americanus, L.

CELASTRACEÆ.

Celastrus scandens, L.
Euonymus atropurpurens, Jacq.   Exterminated.
Euonymus Americanus L., var abovatus, Torr and Gray.

SAPINDACEÆ.

Staphylea trifolia, L.   ·
Acer saccharinum, Wang.
Acer saccharinum var nigrum.   Not common, Prof. M. W.
Harrington.
Acer dasycarpum, Ehrhart.
Acer rubrum, L.

POLYGALACEÆ.

Polygala sanguinea, L.
Polygala verticillata, L.
Polygala Senega, L.   ·
Palygala polygama, Walt.   Have not been seen since 1871.
Polygala pancifolia.   Willd, Tamarack swamps.

SÆGNMINOSÆ.

Lupinus perennis, L.

Trifolium pratense, L.

Trifolium repens, L.

Melilotus officinalis. Willd. Not common.

Melilotus alba, Lam.

Medicago sativa, L.

Medicago lupulina, L.

Amorpha canescens, Nutt. Spec. in University Herb. from Ann Arbor, Prof. M. W. Harrington.

Astragalus Canadensis, L.

Desmodium nudiflorum, D. C.

Desmodium acuminaturn, D. C.

Desmodium rotundifolium, D. C.

Desmodium canescens, D. C. Campus, 1866.

Desmodium cuspidatum, Torr. and Gray.

Desmodium Dillenii, Darlinght. Prof. M. W. Harrington.

Desmodium paniculatum, D. C.

Desmodium Canadense, D. C.

Desmodium rigidum, D. C. Geol. Survey, 1860.

Lespedeza repens, Torr and Gray. Geol. Survey, 1860.

Lespedeza violaceæ, Pers.

Lespedeza violaceæ, var divergens. Prof. M. W. Harrington.

Lespedeza hirta, Ell.

Lespedeza capitata, Michx.

Vicia Cracca, L. Prof. M. W. Harrington.

Vicia Caroliniana, Walt.

Vicia American, Muhl. Geddesburg.

Lathyrus maritimus, Bigelow. Prof. M. W. Harrington.

Lathyrus ochroleucus, Hook.

Lathyrus palustris, L. River bank.

Lathyrus palustris, var myrtifolins. Prof. M. W. Harrington.

Apios tuberosa, Muench.

Amphicarpaea monoica, Nutt.

Baptisia tinctoria. Railroad bridge.

Baptisia leucantha, Torr and Gray.
Cercis Canadensis, L.
Cassia Marilandica, L.
Gymnocladus Canadensis, Lam.

ROSACEÆ.

Prunus Americana, Marshall.
Prunus Pennsylvanica, L.   Prof. M. W. Harrington.
Prunus Virginiana, L.
Prunus serotina, Ehrhart.
Spirea opulifolia, L.
Spirea salicifolia, L.
Poterium Canadense.   Geddesburg.
Agrimonia Eupataria, L.
Geum Album, Gmelin.
Geum Virginianum, L   Geol. Survey, 1860.
Geum strictum, Ait.
Geum rivale, L.
Potentilla Norvegica, L.
Potentilla Canadensis, L.
Potentilla Canadensis, var simplex.   Prof. M. W. Harrington.
Potentilla argentea, L.   Local.
Potentllla arguta, Pursh.
Potentilla Anserina, L.   On Dr. Porter's place.
Potentilla fruticosa, L.
Potentilla palustris, Scop.   In marsh around the lakes West of Ann Arbor.
Fragaria Virginiana, Ehrhart.
Fragaria vesea, L.   Tamarack swamps.
Dalibarda repens, L.   Geol. Survey, 1860.
Rubus triflorus, Richardson.
Rubus strigosus, Michx.
Rubus occidentalis, L.
Rubus Villosus, Ait.
Rubus Canadensis, L.   Prof. M. W. Harrington.
Rubus hispidus, L.   Along Railroad bank.

Roso Carolina, L.
Rosa lucida, Ehrhart.
Rosa Rubiginosa, L.. Road side.
Cratægus coccinea, L.
Cratægus tomentosa, L.
Cratægus tomentosa, var mollis. Prof. M. W. Harrington.
Cratægus tomentosa, var pyrifolia. Prof. M. W. Harrington.
Cratægus tomentosa, var punctata. Prof. M. W. Harrington.
Pyrus coronaria, L.
Pyrus arbutifolia, L.
Pyrus arbutifolia, var melanocarpa. Prof. M. W. Harrington.
Amelanchier Canadensis, Torr and Gray var Botryapium.
Amelanchier Canadensis, Torr and Gray var oblangifolia.

SAXIFRAGACEÆ.

Ribes Cynosbatia, L.
Ribes hirtellum, Michx. Not common.
Ribes floridum, L.
Ribes rubrum, L. Rare.
Parnassia Caroliniana, Michx.
Saxifraga Pennsylvanica L.
Henchera Americana, L.
Mitella diphylla L.
Mitella nuda, L. Tamarack swamp.

CRASSULACEÆ.

Penthorum sedoides, L.

HAMAMELACEÆ.

Hamamelis Virginica, L.

ONAGRACEÆ.

Circaea Lutetiana, L.
Circaea alpina, L.
Epilobium angustifalium, L. On newly cleared land.
Epilobium molle, Torr.

Epilobium coloratum, Muhl.

Oenothera biennis, L.

Oenothera biennis L., var muricata. Miss C. Watson.

Oenothera biennis L., var parviflora. Miss C. Watson.

Oen thera fruticosa, L.

Ludwegia palustris, Ell.

CUCURBITCEÆ.

Echinocystis lobata, Torr and Gray. River bank.

UMBELIFERÆ.

Hydrocotyle Americana, L. Near the first railroad bridge West, in 1861. Dr. A. B. Lyons has since found it near the foundry.

Sanicula Canadensis, L. Rare.

Sanicula Marilandica, L.

Daucus Carota, L. Not common.

Heracleum Canatum, Michx.

Pastinaca sativa, L.

Archemora rigida, D. C.

Archangelica hirsuta, Torr and Gray.

Archangelica atropurpurea, Hoffm. Geol. Survey, 1860.

Coneoselinum Canadense, Torr and Gray.

Thaspium aureum, Nutt.

Zizia integerrima, D. C.

Circuta maculata, L.

Circuta bulbifera, L.

Sium lineare, Michx.

Cryptotænia Canadensis, D. C.

Osmarrhiza longistylis, D. C.

Osmarrhiza brevistylis, D. C.

Erigenia bulbosa, Nutt.

ARALIACEÆ.

Aralia racemosa, L.

Aralia nudicaulis, L.

Aralia quinquefolia. In a ravine two miles' northwest of Ann Arbor.

Aralia trifolia.

CARNACEÆ.

Cornus Canadensis, L.   Tamarack swamps.
Cornus florida, L.
Cornus circinata, L., Herb.   Not common.
Cornus sericea, L.
Cornus stolonifera. Michx.
Cornus paniculata, L., Her.
Cornus alternifolia, L.
Nyssa multiflora, Wang.   Geol. Survey, 1860.

CAPRIFOLIACEÆ.

Lonicera flava, Sims.   Geol. Survey, 1860.
Lonicera parviflora, Lam.
Lonicera parviflora, var Dauglarii.   Not common.
Diervilla trifida Moench.   Geol. Survey, 1860.
Triosteum perfoliatum, L.
Sambucus Canadensis, L.
Sambucus pubens, Michx.
Viburnum Lentaga, L.
Viburnum pubescens, Pursh.
Viburnum acerifolium, L.
Viburnum Opulus, L.   Not common.

RUBIACEÆ.

Galium Aparine, L.
Galium asprellum, Michx.
Galium concinnum, Torr and Gray.
Galium trifidum, L.
Galium triflorum, Michx.
Galium pilosum, Ait.
Galium circæzans, Michx.
Galium lanceolatum, Torr.
Galium bareale, L.
Cephalanthus occidentalis, L.
Mitchella repens, L.   Tamarack swamp.
Honstonia purpurea, L.   Geol. Survey, 1860.

VALERIANACEÆ.

Valeriana sylvatica, Richards.

Valeriana edulis, Nutt.   Not common.

DIPSACEÆ.

Dipsacus sylvestris, Mill.

COMPOSITÆ.

Vernonia fasciculata, Michx.

Liatris squarrosa, Willd.   Prof. M. W. Harrington.

Liatris cylindracea, Michx.

Liatris scariosa, Willd.

Eupatorium purpureum, L.

Eupatorium sessilifolium, L.   Local.

Eupatorium, perfoliatum, L.

Eupatorium, ageratoides, L.

Aster macrophyllus, L.

Aster patens var phlogifolins.   Geol. Survey, 1860.

Aster lævis, L.   Geol. Survey, 1860.

Aster lævis var lævigatus.   Geol. Survey, 1860.

Aster lævis var cyaneus.

Aster azureus, Lindl.

Aster undulatus, L.

Aster cordifolius, L.

Aster sagittifolius, L.

Aster multiflorus, Ait.

Aster miser.   Prof. M. W. Harrington.

Aster longifolius, Lam.   Geol. Survey, 1860.

Aster puniceus, L.

Aster Novæ-Angliæ, L.

Erigeron Canadense, L.

Erigeron bellidifolium, Muhl.

Erigeron Philadelphicum, L.

Erigeron annum, Pers.

Erigeron Strigosum, Muhl.

Diplopappus umbellatus, Torr and Gray.

Solidago latifolia, L.

7

Solidago cæsia, L.
Solidago speciosa, Nutt. Geol. Survey, 2860.
Solidago speciosa var angustata. Geol. Survey, 1890.
Solidago rigida, L.
Solidago Riddellii, Frank.
Solidago patula, Muhl. Geol. Survey, 1860.
Solidago arguta, Ait. Geol. Survey, 1860.
Solidago arguta var scabeella. Geol. Survey, 1860.
Solidago altissima, L.
Solidago memoralis, Ait. Prof. M. W. Harrington.
Solidago Canadensis, L.
Inula Helennum, L.
Polymnia Canadensis, L. Local.
Polymnia Uvedalia, L. Prof. M. W. Harrington.
Silphium terebinthinaceum, L.
Ambrosia trifida, L.
Ambrosia trifida, var integrifolia.
Ambrosia artemissæfolia, L.
Xanthium strumarium, L.
Heliopsis lævis Pers.
Heliopsis lævis, var scabra.
Rudbeckia lacniata, L.
Rudbeckia speciosa, Wenderoth. Geol. Survey, 1860.
Rudbeckia fulgida, Ait. Geol. Survey, 1860.
Helianthus occidentalis, Riddell.
Helianthus gigantens, L.
Helianthus strumosus, L.
Helianthus divaricatus, L.
Helianthus hirsutus, Raf.
Helianthus decapetalus, L.
Helianthus doronicoides, Lam. Geol. Survey, 1860.
Coreopsis tripteris, L.
Coreopsis aristosa, Michx.
Bidens frondosa, L.
Bidens cerunu, L.
Bidens chrysanthemoides, Michx.
Bidens Bechii, Torr. Huron River, Ann Arbor.

Helenium Autumnale, L.

Maruta Cotula, D. C.

Achillea Millefolium, L.

Leucanthemum vulgare, Lam.

Leucanthemum Parthenium, Godran. In the streets, escaped from cultivation.

Tanacetum vulgare, L.

Artemisia biennis, Willd.

Gnaphalium uliginosum, L.

Gnaphalium polycephalum, Michx.

Antennaria plantaginifolia, Hook.

Erechthites hieracifolia, Raf.

Cacalia atriplicinifolia, L.

Senecia aureus, L.

Senecia aureus, var obavatus. Geol. Survey, 1860.

Senecia balsamitæ.

Cirsium lanceolatum, Scop.

Cirsium discolor, Spreng.

Cirsium muticum, Michx.

Cirsium pumilum, Spreng.

Cirsium arvense, Scop.

Cirsium altissimum, Spreng.

Lappa officinalis, Allioni, var majas.

Cichorium Intybus, L.

Cynthia Virginica, Dov.

Hieracium Canadense, Michx.

Hieracium scabrum, Michx.

Hieracium venosum, L.

Nabalus albus, Hook.

Nabalus albus, var serpentaria. Geol. Survey, 1860.

Nabalus altissimus, Hook. Geol. Survey, 1860.

Taraxacum Dens—leonis, Desf.

Lactuca Canadensis, L.

Lactuca Canadensis, var integrifolia.

Sonchus oleracus, L.

Sonchus asper, Vill.

LOBELIACEÆ.

Lobelia cardinalis, L.
Lobelia syphilitica, L.
Lobelia spicata, Lam.
Lobelia Kalmii.

CAMPANULACEÆ.

Campanula rotundifolia, L.
Campanula aparinoides, Pursh.
Campanula Americana, L.

ERICACEÆ.

Gaylussacia resimosa, Torr and Gray.
Gaylussacia frondosa, Torr and Gray.   Geol. Survey, 1860.
Vaccinnium Oxycoccus, L.
Vaccinnium macrocarpon, Ait.
Vaccinnium Pennsylvanicum, Lam.   Prof. M. W. Harrington.
Vaccinnium vacillans, Solander.   Prof. M. W. Harrington.
Vaccinnium Canadense, Kalm.   Prof. M. W. Harrington.
Chiogenes hispidula, Torr and Gray.   Local.
Cassandra calyculata, Dan.   Local.
Andromeda polifolia, L.
Pyrola rotundifolia, L.
Pyrola elliptica, Nutt.
Pyrola secunda, L.
Chimaphila umbellata, Nutt.
Monotropa uniflora, L.
Monotropa Hypopitys, L.   Rare in vicinity of Ann Arbor.

AQUIFOLIACEÆ.

Ilex verticllata, Gray.

PLANTAGINACEÆ.

Plantago Major, L.
Plantago lanceolata, L.

PRIMULACEÆ.

Trientalis Americana, Pursh.
Lysimachia thyrsiflora, L.

Lysimachia stricta, Ait.
Lysimachia quadrifolia, L.
Lysimachia ciliata, L.
Lysimachia longifolia, Pursh.
Anagallis arvensis, L.   Geol. sur. 1860.

LENTIBULACEÆ.

Utricularia vulgaris, L.
Utricularia minor, L.   Geol. sur. 1860.
Utricularia intermedia, Hayne.   Geol. sur. 1860.

OROBANCHACEÆ.

Epiphegus Virginiana, Bart.   Rare in vicinity of Ann Arbor.
Conopholis Americana, Wallroth.   Geol. sur. 1860.
Aphyllon uniflorum, Torr and Gray.

SCROPHULARIACEÆ.

Verbascum Thapsus, L.
Verbascum Blattaria, L.   Prof. M. W. Harrington.
Linaria vulgaris, Mill.
Scrophularia nodosa, L.
Collinsia verna, Nutt.   Lost.
Chelone glabra, L.
Pentstemon pubescens, Solander.
Mimulus ringens, L.
Ilysanthes gratioloides, Benth.
Veronica Virginica, L.
Veronica Anagallis, L.
Veronica Americana, Schweinitz.
Veronica scutellatta, L.
Veronica officinalis, L.
Veronica serpyllifolia, L.
Veronica peregrina, L.
Veronica arvensis, L.
Veronica agrestis, L.   Prof. M. W. Harrington.
Gerardia tenuifolia, Vahl.
Gerardia flava, L.   Prof. M. W. Harrington.
Gerardia quercifolia, Pursh.   Prof. M. W. Harrington.

Gerardia auriculata, Michx.   Prof. M. W. Harrington.
Gerardia pedicularia, L.
Castilleia coccinea, Spreng.
Pedicularis Canadensis, L.
Pedicularis lanceolata, Michx.

ACANTHACEÆ.

Dianthera Americana, L.

VERBENACEÆ.

Verbena hastata.
Verbena urticifolia, L.
Phyrma Leptostachya, L.

LABIATÆ.

Teuchrium Canadense, L.

Mentha viridis, L.

Mentha peperita, L.

Mentha Canadensis, L.

Lycopus Virginicus, L.

Lycopus Europæns, L.

Pycnanthemum lanceolatum, Pursh.

Pycnanthemum linifolium, Push.   Geol. sur. 1860.

Hedeoma pulegroides, Pers.   .

Colinsonia Canadensis, L.

Monarda fistulosa, L.

Blephalia ciliata, Raf.

Lophanthus scrophulariæfolius, Benth.

Nepeta Cataria, L.

Nepeta Glechoma, Benth.

Physostegia Virginiana, Benth.   Geol. sur. 1860.

Brunella vulgaris, L.

Sentellaria galericulata, L.

Sentellaria lateriflora, L.

Stachys palustris var. aspera.

Leonurus Cardiaca, L.

BARAGINACEÆ.

Symphytum officinale, M.   Sparingly escaped from cultivation, M. W. Harrington.

Lithospermum arvense, L.
Lithospermum latifolium, Michx.
Lithospermum canescens, Lehm.
Myasotis verna, Nutt.   Geot. sur. 1860.
Echinospermum officinale, L.
Cynoglossum Morisoni, D. C.

HYDROPHYLLACEÆ.

Hydrophyllum Virginicum, L.
Hydrophyllum Canadense, L.
Hydrophyllum appendiculatum, Michx.

PALEMONIACEÆ.

Phlox pilosa, L.
Phlox divaricata, L.

CONVOLONLACIÆ.

Calystegia sepium, R. Br.
Calystegia spithamæa, Pursh.
Cuscuta Gronovii, Willd.

SALANACEÆ.

Solanum Dulcamora, L.
Solanum nigrum, L.
Physalis pubescens, L.   Prof. M. W. Harrington.
Physalis viscosa, L.   Prof. M. W. Harrington.
Nicandra physaloides, Gærtu.
Datura Stramonium, L.
Datura Tatula, L.

GENTIANACEÆ.

Gentiana quinqueflora, Lam.
Gentiana quinqueflora, var. occidentalis.   Geol. sur. 1860.
Gentiana crinita, Froel.
Gentiana detonsa, Fries.   Prof. M. W. Harrington.
Gentiana alba, Muhl.
Gentiana Andrewsii, Grisel.
Gentiana puberula, Michx.
Menyanthes trifoliata, L.

APOCYNACEÆ.

Apocynum androsæmifolium, L.

Apocynum cannabinum, L.

ASCLEPIADACEÆ.

Asclepias Cornuti, Decaisne.

Asclepias variegata, L. Geol. sur. 1860.

Asclepias phytolaccoides, Pursh.

Asclepias quadrifolia, Jacq. Geol. sur. 1860.

Asclepias purpurascens, L.

Asclepias incarnata, L.

Asclepias tuberosa, L.

Asclepias verticillata, L.

Acerates viridiflora, Ell.

OLEACEÆ.

Fraxinus Americana, L.

Fraxinus viridis, Michx. Geol. sur. 1860.

Fraxinus sambucifolia, Lam. Geol. sur. 1860.

ARISTOLOCHIACEÆ.

Asarum Canadense, L.

CHENOPODIACEÆ.

Chenopodium album.

Chenopodium hybridum, L.

Chenopodium Botrys, L. Geol. sur. 1560.

Chenopodium ambrosioides, L. Prof. M. W. Harrington.

AMARANTACEÆ.

Amarantus retroflexus, L. var. hybridus. Prof. M. W. Harrington.

Amarantus albus, L. Prof. M. W. Harrington.

Amarantus hypochondriacus, L. Geol. sur. 1860.

POLGONACEÆ.

Polygonum orientale, L.

Polygonum incarnatum, Ell.

Polygonum Persicaria.

· Polygonum Hydropiper, L.

Polygonum acre, H. B. K.
Polygonum hydropiperoides, Michx.
Polygonum amphibium, L.
Polygonum Virginianum, L.
Polygonum aviculare, L.
Polygonum aviculare, var. erectum, Roth.
Polygonum tenne, Michx.
Polygonum sagittatum, L.
Polygonum Convolvulus, L.
Polygonum dumetorum, L. var. scandens.

Fagopyrum esculentum, Moench. Escaped from cultivation.

Rumex verticillatus, L. Geol. sur. 1860.
Rumex crispus, L. Prof. M. W. Harrington.
Rumex obtusifolius, L.
Rumex sanguineus, L Geol. sur. Geol sur. 1860.
Rumex Acetosella, L.

LAURACEÆ.

Sassafras officinale, Neis.
Lindera Benzoin, Meisner.

THYMELEACEÆ.
Dirca palustris, L.
ELEAGECEÆ.

Shepherdia Canadensis, Nutt. Lost.

SANTALACEÆ.

Comanda umbellata, Nutt.

SANRURACEÆ.

Saururus cernuus, L. Local.

EUPHORBIACEÆ.

Euphorbia maculata, L.
Euphorbia hypericifolia, L.
Euphorbia corollata, L.
Euphorbia Esula, L. Sparingly escaped from cultivation.
M. W. Harrington.

Euphorbia commutata, Engelm. Geol. sur. 1860.

Euphorbia Cyparissias, L. Escaped. Common in the city, M. W. Harrington.

URTRICACEÆ.

Ulmus fulva, Mich.

Ulmus Americana, L.

Ulmus racemosa, Thomas. Swamp one mile south of Ann Arbor. Geol. sur. 1860.

Morus Alba, L.

Urtica gracilis, Ait.

Laportea Canadensis, Gaudichand.

Pilea pumila, Gray.

Bochmeria cylindrica, Willd.

Cannabis Sativa, L.

Humulus Lupulus, L.

PLATANACEÆ.

Platanus occidentalis, L.

JUGLANDACEÆ.

Juglans cinerea, L.

Juglans nigra, L.

Carya alba, Nutt.

Carya microcarpa, Nutt. Prof. M. W. Harrington.

Carya porcina, Nutt. Prof. M. W. Harrington.

Carya amara, Nutt. Prof. M. W. Harrington.

Carya Sulcata, Nutt. Geol. sur. 1860.

CUPULIFERÆ.

Quercus alba, L.

Quercus bicolor, Willd. Prof. M. W, Harrington.

Quercus macrocarpa, Michx.

Quercus Prinus, L. var. acuminata Michx.

Quercus imbricaria, Michx.

Quercus coccinea, Wang.

Fagus ferruginea, Ait.

Corylus Americana, Walt.

Ostrya Virginica, Willd.

Carpinus Americana, Michx.

BETULACEÆ.

Betula lenta, L.   Tamarack swamp.
Betula alba var. populifolia, Spach.
Betula pumila, L.

SALICACEÆ.

Salix discolor, Muhl.   M. W. Harrington.
Salix Cordata, Muhl., var. angustata.
Salix livida, Wahl, var. occidentalis.
Salix fragilis, L.
Salix nigra, Marsh.   M. W. Harrington.
Salix lucida, Muhl.
Populus tremuloides, Michx
Populus grandidentata, Michx.
Populus balsamifera, L., var. candicans.
Populus Alba.

CONIFERA.

Larix Americana, Michx.   Tamarack swamp.
Juniperus communis, L.
Juniperus Virginiana, L.

ARACEÆ.

Arisæma triphyllum, Torr.
Arisæma Dracontium, Schott.
Peltandra Virginica, Raf.   Huron River.
Calla palustris, L.
Symplocarpus fœtidus, Salisb.
Acorus Calamus, L.   Huron River.

LEMNACEÆ.

Lemna trisulca, L.   Ponds in Cemetery, M. W. Harrington.
Lemna minor, L.   Ponds in Cemetery.
Lemna polyrrhiza, L.   Ponds in Cemetery.

TYPHACEÆ.

Typha latifolia, L.
Sparganium eurycarpum, Engelm.
Sparganium simplex, Hudson, var. androcladum.

NAIADACEÆ.

Naias flexilis, Rostk. Huron River.
Potamogeton natans, L, Huron River.
Potamogeton perfoliatus, L. Huron River.
Potamogeton pectinatus, L. Huron River.

ALISMACEÆ.

Triglochin maritmum, L. var. elatum.
Alisma Plantago, L. var. Americanum.
Sagittaria variabilis, Engelm.

HYDROCHARIDACEÆ.

Anacharis Canadensis.

ORCHIDACEÆ.

Orchis spectabilis, L. Not common.
Habenaria tridentata, Hook.
Habenaria virescens, Spreng.
Habenaria viridis, R. Br., var. bracteata, Reichenbach.
Habenaria hyperborea, R. Br.
Habenaria dilatata, Gray.
Habenaria Hookeri, Torr.
Habenaria ciliaris, R. B.
Habenaria Leucophaea.
Habenaria lacera, R. Br.
Habenaria psycodes, Gray.
Habenaria fimbriata, R. Br. Geol. sur.
Spiranthes latifolia, Torr.
Spiranthes cernua, Richard.
Spiranthes gracilis, Bigelow.
Arethusa bulbosa, L. Peat bogs two miles west of Ann Arbor.
Pogonia ophioglossoides, Nutt.
Calopagon pulchellus, R. Br.
Microstylis ophioglossoides, Nutt.
Liparis Lœselii, Richard.
Corallarhiza liliifolia multiflora, Nutt.
Aplectrum hyamale, Nutt.
Cypripedium candidum, Muhl.

Cypripedium parviflorum, Sisb.   M. W. Harrington.
Cypripedium pubescens, Willd.
Cypripedium spectabile, Swartz.
Cypripedium acaule, Ait.   Tamarack swamp.

AMARYLLIDACEÆ.

Hypoxys erecta, L.

HÆMODARACEÆ.

Aletris farinosa, L.

IRIDACCEÆ.

Iris versicolor, L.
Sisyrinchium Bermudiana, L.

DISCOREACEÆ.

Discorea villosa, L.

SMILACEÆ.

Smilax rotundifolia, L.
Smilax hipida, Muhl.   Geol. sur. 1860.
Smilax herbacea, L.
Smilax herbacea, var. pulverulenta.
Smilax tamnifolia, Michx.   Geol. sur. 1860.

LILIACEÆ.

Trillium grandiflorum, Salisb.
Trillium erectum, L.
Trillium erectum, var. album.
Trillium erectum, var. declinatum.   M. W. Harrington.
Trillium erythocarpum, Michx.   M. W. Harrington.
Zygadenus glaucus, Nutt.   Rare.
Tofieldia glutinosa, Willd.
Uvularia grandiflora, Smith.
Uvularia perfoliata, L.   M. W. Harrington.
Uvularia sessilifolia, L.
Smilacina racemosa, Desf.
Smilacina stellata, Desf.
Smilacina bifolia, Ker.
Polygonatum biflorum, Ell.
Polygonatum giganteum, Dietrich.

Lilium Philadelphicum, L.
Lilium Canadense, L.
Lilium superbum, L.   Geol. sur.
Erythronium Americanum, Smith.
Erythronium albidum, Nutt.
Allium tricoccum, Ait.
Allium cernuum, Roth.
Allium Canadense, Kalm.   Local.

JUNCACEÆ.

Luzula comprestis, D. C.
Juncus effusus, L.
Juncus bufonis. L.
Juncus nodosus, L.
Juncus, tenuis, Willd.
Juncus Pelocarpus, E. Meyer.   M. W. Harrington.
Juncus acuminatus, Michx, var. legitimus.   M. W. Harrington.
Juncus Canadensis, Gray.

PONTEDERIACEÆ.

Pontederia cordata, L.
Schollera graminea, Willd.

COMNELYNACEÆ.

Tradescantia Virginica, L.

CYPERACEÆ.

Cyperus flavescens, L.   M. W. Harrington.
Cyperus diandrus, Torr.
Cyperus Strigosus, L.   M. W. Harrington.
Cyperus filiculmis, Vahl.
Dulchium spathaceum, Pers.
Elocharis obtusa, Schultes.   M. W. Harrington.
Elocharis palustris, R. Br.
Elocharis tenuis, Schultes.
Elocharis acicularis, R. Br.
Scirpus pungens, Vahl.

Scirpus validus, Vahl.

Scirpus atrovirens, Muhl.

Scirpus polyphyllus, Vahl.

Scirpus lineatus, Michx.

Scirpus Eriphorum, Michx.

Eriophorum vaginatum, L.   M. W. Harrington.

Eriophorum Virginicum, L.

Eriophorum polystachyon, L.

Eriophorum polystachyon, var. augustifolia.   M. W. Harrington.

Eriophorum gracile, Koch.

Fimbristilis autumnalis, Roem and Schult.   M. W. Harrington.

Fimbristilis capillaris, Gray.   M. W. Harrington.

Rynchaspora alba, Vahl.

Scleria triglomerata, Michx.   M. W. Harrington.

Carex polytrichoides, Muhl.

Carex teretinscula, Good.

Carex decomposita, Muhl.

Carex vulpinoidea, Michx.

Carex stipata.

Carex sparganoides, Muhl.

Carex cephaloides, Dew.

Carex cepalophara, Muhl.

Carex straminea, Schk.

Carex straminea, var. typica.   M. W. Hrrrington.

Carex straminea, var. tenera.   M. W. Harrington.

Carex stricta, Lam

Carex crinita, Lam.

Carex limosa, L.

Carex aurea, Nutt.   .

Carex rosea, Schk.   Prof. M. Harrington.

Carex granularis, Muhl.   Prof. M. W. Harrington.

Carex conoidea, Schk.   M. W. Harrington.

Carex conescens, L.   M. W Harrington.

Carex scoparia, Schk.   M. W. Harrington.

Carex lanuginosa, Michx. M. W. Harrington.
Carex scabrata. Schw. M. W. Harrington.
Carex gracillima, Schw.
Carex laxiflora, Lam.
Carex Pennsylvanica, Lam.
Carex comosa, Booth
Carex hystricina, Will.
Carex instumescens, Rudge.
Carex lupulina, Muhl.
Carex Tuckermania, Booth.

GRAMINEÆ.

Leersia oryzoides, Swartz.
Leersia Virginica, Willd.
Zizania aquatica, L.
Alapecures pratensis, L.
Alapecurus aristulatus, Michx.
Phleum pratense, L.
Agrostis perennans, Tuckerm. M. W. Harrington.
Arostis scabra, Willd. M. W. Harrington.
Agrostis vulgaris, With. M. W. Harrington.
Muhlenbergia diffusa, Schreber.
Brachyelytrum aristatum. M. W. Harrington.
Calamagrostis Canadensis, Willd.
Spartina cynosuroides. Willd.
Dactylis glomerata, L.
Kœleria cristata, Pers. M. W. Harrington.
Eatonia obtusata, Gray. M. W. Harrington.
Eatonia Pennsylvanica, Gray. M. W. Harrington.
Glyceria elongata, Trin. M. W. Harrington.
Glyceria nervata, Trin. M. W. Harrington.
Glyceria aquatica, Smith.
Poa annua, L. M. W. Harrington.
Poa compressa, L. M. W. Harrington.
Poa serotina, Ehrhart. M. W. Harrington.
Poa pratensis, L. M. W. Harrington.
Eragrostis poæoides, Beauv. var. megastachya.

Eragrostis capillaris, Nees. M. W. Harrington.
Festuca tenella, Willd. M. W. Harrington.
Festuca ovina, L. M. W. Harrington.
Festuca Elatior, L. var, pratensis. M. W. Harrington.
Festuca nutans, Willd. M. W. Harrington.
Bromus secalinus, L.
Bromus ciliatus, L.
Phragmites communis, Trin.
Triticum repens, L.
Elymus Virginicus, L. M. W. Harrington.
Elymus Canadensis, L.
Gymnostichum Hystrix, Screb.
Danthania spicata, Beauv. M. W. Harrintgon.
Avena striata, Michx. M. W. Harrington.
Aira cæspitosa, L. M. W. Harrington.
Hierochloa borealis, Reom and Schultes. Huron river.
Phalaris Canadensis, L.
Phalaris arundinacea, L. M. W. Harrington.
Philaris arundinacea, var picta. M. W. Harrington.
Panicum sanguinale, L.
Panicum capillare, L.
Panicum Crus-galli, L.
Panicum latifolium, L. M. W. Harrington.
Panicum dichotomum, L. M. W. Harrington.
Panicum depauperatum, Muhl. M. W. Harrington.
Panicum glabrum, Gaudin. M. W. Harrington.
Setaria glauca, Beauv.
Setaria viridis, Beauv.
Cenchrus tribuloides, L.
Andropagun furcatus, Muhl.
Andropagun scoparius, Michx.
Sorghum nutans, Gray.

EQUISETACEÆ.

Equisetum arvense, L.
Equisetum sylvaticum, L. Rare.
Equisetum limosum L. M. W. Harrington.
Equisetum hyemale, L.

8

Adiantum pedatum, L.
Pteris aquilina, L.
Woodwardia Virginica, Smith.
Asplenium angustifolium, Michx.   Rare about Ann Arbor.
Asplenium thelypteroides, Michx.
Asplenium Filix-fœnima, Bernh.
Phegopteris hexagonoptera, Fee.
Aspidium Thelypteris, Swartz.
Aspidium Noveboracense, Swartz.
Aspidium spinulosum var intermedium.
Aspidium spinulosum var dilatatum.
Aspidium cristatum, var Clintonianum.
Aspidium acrostichoides, Swartz.
Cystopteris bulbifera, Bernh.
Cystopteris fragilis, Bernh.
Struthiopteris Germanica, Willd.
Onoclea sensibilis, L.
Osmunda regalis, L.
Osmunda Claytoniana, L.
Osmunda cinamomea, L.
Botrychium Virginicum, Swartz.
Botrychium lunaroides, Swartz.   Rare.

LYCOPADIACEÆ.

Salaginella apus, Spring.

EXTERMINATED PLANTS.

Ranunculus multifidus, Pursh.
Dicentra Cucullaria, D. C.
Arabis Drommandii, Gray.
Erodium circutarium, L., Herb.
Polygala polygama, Walt.
Nyssa multiflora, Wang.
Aphyllon uniflorum, Torr and Gray.
Collinsia verna, Nutt.
Dirca palustris, L.

Calla palustris, L.
Liparis liliifolia, Richard.
Liparis Loeselii.
Botrychium lunarioides, Swartz.

RARE AND LOCAL PLANTS.

Thalictrum purpurascens, L.
Captis trifolia, Salisb.
Liriodendron Tulipifera, L.
Asimina triloba, Dunal.
Jefferzonia diphylla, Pers.
Brasenia peltata, Pursh.
Solea concolor, Ging.
Viola rostrata, Pursh.
Viola striata, Ait.
Viola Canadensis, L.
Elodes Virginica. Nutt.
Rhus Aromatica, L.
Poterium Canadense.
Hydracotyle Americana, L.
Sanicula Canadensis, L.
Erigenia bulbosa, Nutt.
Aralia quinguefolia.
Bidens Beckii, Torr.
Chimaphila, umbellata, Nutt.
Monotrapa Hypopitys, L.
Epiphegus Virginiana, Bart.
Calystegia spithamaea, Pursh.
Shepherdia Canadensis, Nutt.
Saururus cernuris, L.
Peltandra Virginica, Raf.
Spiranthes latifolia, Torr.
Corallorhiga multiflora.
Aplectrum hyamale, Nutt.
Cypripedium Candidum, Muhl.
Allium Canadense, Kalm.

Equisetum sylvaticum, L.
Asplenium angustifolium, Michx.

INTRODUCED PLANTS.

Ranunculus bulbosus, L.
Ranunculus acris, L.
Nasturtium officinale, R. Br.
Thlaspi arvense, L.
Hypericum pyramidatum, Ait.
Linum Virginianum, L.
Medicago lupulina.
Leucanthemum vulgare, Lam.
Cirsium arvense, Scop.
Virbascum Blattaria, L.
Veronica agrestis, L.
Symphytum officinale, L.
Datura.
Euphorbia Esula, L.
Euphorbia Cyparissisas, L.
Cenchrus tribuloides, L.

H.

COLORED SNOW-FALL IN WESTERN MICHIGAN.

BY S. T. DOUGLAS.

On the 5th of February last there occurred in the western part of Michigan a somewhat strange and interesting phenomenon, namely, a fall of colored snow, which, although exciting much interest and some speculation at the time, has as yet met with no very satisfactory explanation which would account for its origin and its appearance. Coming into possession of a specimen of the sediment obtained by the evaporation of the snow, I became somewhat interested in the subject, and made some inquiries which, together with some subsequent work, which owing to the lack of material and of sources of information is very incomplete, I venture to present.

To attempt to maintain any particular theory which may account for this phenomenon, would doubtless be a difficult task in our hands; but by giving the facts concerning it, and comparing it with events of a similar nature which have been recorded, we may perhaps be able to draw some conclusions as to its probable, or at least possible, origin.

The specimens of the sediment obtained from the snow by evaporation were sent, one by Mr. J. G. Williams, of Saugatuck, Allegan County, Mich., and the other, which was received later, from H. D. Post, Esq., of Holland, Ottawa County, The sediment obtained from Saugatuck was secured by the evaporation of about two quarts of the snow. It is in the form of an impalpable powder, of a grayish black color, presenting somewhat the appearance of emery, and containing a varying amount of fibrous material.

From Mr. Williams's letter the following information is gained in regard to the circumstances existing at the time of the occurrence :

" The wind for two or three days previous to the 5th of February had been blowing strongly from the west, but during the day the weather had been pleasant, and at the time the snow fell there was scarcely any wind—what there was being from the south-west. The ground was already covered with snow, from a previous fall. Between the hours of four and six P. M. the large flakes of light snow were noticed to present a strange appearance, coloring the snow as they fell. This continued until it had reached a depth of three or four inches, when the wind came strong, blowing quite a gale from the south-west. The snow then came down clear again."

Some of this last fall of snow was gathered, as was stated, and evaporated carefully in a suitable place, yielding the sediment which is had for analysis. These are the circumstances attending the fall of snow at Saugatuck.

The sediment received from Holland resembles very much that from the former place, both in color and in structure, and the condition of the weather and wind at the time was nearly the same. During the fall of this snow at Holland, the atmosphere was noticed to be sensibly warmer ; at one locality the thermometer rising eight degrees during the half hour the snow was falling, falling again after it had ceased and the clear snow had again commenced. The wind was very light during the fall, and from the south-west. Not only at Holland and Saugatuck was this phenomenon observed, but according to both of these informants it extended over a wide range of territory. In Michigan it was noticed from about five miles south of Grand Haven, in Ottawa County, to Ganges, south about thirty miles, in Allegan County. Inland it was noticed at Allegan and other places. It must, therefore, have fallen over an extent of from three to six hundred square miles in Michigan, after crossing Lake Michigan.

Our informant also vouches for the statement that the same phenomenon was noticed in Illinois, sixty miles south-west of Chicago, and also at Janesville, Wis. The wind, at both of these places, being in the same direction as in Michigan.

In order to confirm this latter statement, I have written to the Signal Service Bureaus at different points in Illinois, Missouri and Wisconsin, but as yet no reply has been received.

That some idea may be formed as to the amount of this sediment which fell, a square yard was measured off at Holland soon after the storm had ceased, and after evaporation the upper layer of snow yielded 87 grains of solid matter. Assuming that the fall took place over an extent of 400 square miles, and that it was equal over the whole, we may easily calculate that with this fall of snow there were precipitated about 7,000 tons of solid or earthy matter.

Having now the true facts of the case, it remains to account for this, to say the least, singular phenomenon, in some rational way. It is evident, both from appearance and from analysis, that the sediment from either locality is identical in almost every respect, and is without doubt of the same origin ; both have the same color ; both contain fibrous matter ; and a description of the one will to a great extent answer for the other. Under the microscope the substance shows itself to be composed of organic and inorganic matter. The inorganic portion apparently consists of small particles of silica, some of which seem to be colored with iron and earthy matter. These particles of silica have not the appearance of having been water worn, but the fracture of them is sharp and well defined, having all the appearance of being recently broken or reduced to powder, without the abrasive action of water. There are also to be seen fine threads or fibres, perfectly transparent, and having a resemblance to fine tubes of glass or silica. These tubes present no appearance of organic structure, and as a whole, nothing either of a vegetable or animal structure is to be detected. Small particles of organic matter.

also, seem to be distributed through the mass. After ignition, the substance turns reddish brown, this color probably being due to ferric oxide. The organic matter is burnt off, and with the exception of the fibrous material, it presents about the same appearance as before.

A qualitative analysis shows the presence of silica, iron, lime, manganese, and slight traces of sulphur. Quantitatively, the sediment from Saugatuck gives 14.7 per cent. of carbonaceous matter and moisture, leaving a residue of 85.3 per cent.,—9.9 per cent. of which is soluble in acids, and 75.4 per cent. insoluble. The soluble portion consists mainly of iron, the other elements existing in very small traces.

The analysis of the sediment from Holland yields a very similar result, it consisting of 16 per cent. of carbonaceous matter and moisture, leaving a residue, 9 per cent. of which is soluble in acids, and 75 per cent. insoluble.

The specific gravity of the two specimens is very nearly the same ; one being 2.063, and the other a trifle less—the diminution in specific gravity of the latter (that from Saugatuck) being easily attributable to a slight accident in manipulation.

It is very evident that in presenting all the circumstances and facts which can in any way bear upon the case, that the meteorological condition of the atmosphere, the direction and velocity of the wind, and perhaps the condition of the surface, as to previous falls of snow, all play an important part in its consideration.

The meteorological observations and reports taken and received at any of the numerous signal stations, certainly furnish all that could be desired on this point. These reports taken at Detroit for Feb. 5th, at 7:11 A. M., and at 4:11 and 10:30 P. M., Detroit time, have been obtained through the kindness of the signal sergeant at Detroit. From what has been said in regard to the state of the wind at Holland and Saugatuck, it being according both to the informants and to the weather reports from W. S. W., it is very evident that whether we consider the material under consideration, whatever it may be, to have been borne by the wind, sweeping along the surface of the earth, or whether it was carried along by the upper current, which moves so steadily from the southwest, it doubtless originated in that direction.

It will, therefore, be somewhat unnecessary to consider the state of the weather at the time, at any other points than those lying in a westerly, southwesterly, or southerly direction.

Starting, then, at the nearest point where reports are taken, which is Grand Haven, at 4 o'clock, the time of the observation, there was a heavy snow falling, and the wind, according to the scale adopted by the weather bureau, was high, having a velocity of 30 miles per hour. An hour later, however, during the fall of the colored snow, it had decreased very much in velocity, and at 10 P. M., the time of the next observation, it had a velocity of 18 miles an hour.

Crossing Lake Michigan to the west and southwest, Chicago and Milwaukee would be the next points of observation. At the former place the wind was southwest, at the rate of 18 miles per hour, weather cloudy. At the latter the wind was from the west, at the rate of 17 miles per hour, and a heavy storm falling. At St. Louis the wind was brisk from the south, with a velocity of 16 miles an hour, and the weather clear. Proceeding farther to the west and southwest, Fort Gibson, in the Indian Territory, seems to be the only place from which observations are reported. At this point there was a fresh wind from the southwest, with a velocity of 10 miles per hour, and the weather clear. These reports are all from the observations taken at 4 P. M. Observations taken at the same points in the morning, show the wind to have been of very much less velocity, in no case being more than 12 or 15 miles per hour. In other words, at no point west, southwest or south was there a wind of any great force.

The weather reports for the few days preceding the 5th are taken from the observations made at Chicago on the 4th. In lower Missouri, Ohio, Tennesee and the Northwest, the barometer was stationary, southwest wind, and cold and clear weather; in the lake region, westerly winds, rising barometer and local snow. On the third, the wind was quite high at Chicago, from the southwest. At points south and west there seems to have been quite a gale, with clear weather and very cold, with previous falls of snow. The full reports were not obtained from Chicago, and the force of the wind is not, therefore, known.

Assuming that we now have the main facts and circumstances which relate to the phenomenon, it remains to be accounted for by some rational theory. What the true theory is, it will perhaps be impossible to say; but those which will naturally present themselves, are first, Could the dust have its source in some neighboring manufacturing establishment, and have escaped from some high chimney or smoke stack?

A second theory might perhaps suggest a meteoric origin.

A third would say that far enough southwest to find dry prairie soil, neither frozen nor covered with snow, a powerful tornado or whirlwind occurred, of sufficient magnitude to carry this great quantity of dust high enough into the air to meet the anti-trade wind, and crossing Lake Michigan, it was deposited with the heavy snow storm which fell in that region.

Again, could the dust in question be derived from Chicago?

Finally, the sediment may be of volcanic origin, borne by the force of volcanic eruption to the upper current, and thus carried to its place of deposit. These are the more likely theories which are to be advanced, and the derivation of the substance is probably due to one of these causes.

Before going into any discussion of them, however, it will be well to notice that this is by no means an isolated case. Phenomena of an analogous nature are very numerous, and have been the occasion of a good deal of investigation and inquiry. By stating some of the more important of these cases, and tracing a similarity, if possible, between them and the present occurrence, we may perhaps arrive at a more definite conclusion as to its true cause. It would be quite an error to confound the phenomenon which we are now considering, with any of these prodigies, a great number of which are collected and noticed in Flammarian's work on the atmosphere; such as showers of sulphur, plants, frogs, fish, etc., all of which are well authenticated. Nor would it be right to class it with the red snow or "*uredo nivalis,*" a kind of microscopic infusoria, which is found more especially in the polar regions. The entire absence under the microscope of any organic form, together with its chemical composition, will certainly preclude any such idea.

We must, therefore, trace it to some other cause. As early as Homer's time, showers of blood were of comparatively frequent occurrence. Flammarian notes many miracles of this kind to have taken place.

In 1744 there fell a red rain in Geneva which terrified the inhabitants, but it was subsequently ascertained that this tint was due to some red earth which a strong wind had carried into the air from a neighboring mountain. In 1608 one of these pretended showers of blood fell at Aix (Provence) which the priests attributed to diabolical influence. This prodigy was, however, examined into very minutely, and what seemed to be red rain, in reality was the excrements of butterflies,* which had been noticed in large numbers. Generally speaking, showers of blood were not only red spots produced by certain insects, but regular showers which the wind had carried into the air. The general origin, according to Flammarian, was not ascertained until the present century.

In 1813 one of these strange red showers fell in the kingdom of Naples and in the two Calabrias. This was examined and analyzed, and a report made before the Naples Academy of Science. An east wind had been blowing for two days, when a dense cloud was noticed moving toward the sea. At two P. M. the sea became calm, but the cloud covered the neighboring mountains, and began to intercept the sun's light. The town was plunged into profound darkness, the storm was very great, and the drops of rain were colored red. The dust gathered was of a yellowish hue, and contained small, hard bodies resembling pyroxene. Heat turned the substance brown. Its specific grav-

---

*Flammarian, p. 151.

ity was 2.07.   Silica, alumina, lime and iron were its chief constituents.*

It was not until 1846 that a general examination of these
rains was made, and their origin found by following them into
space.   On May 16th of that year, an earthy rain fouled the
the waters at Siam.   In the autumn of the same year, there was
a similar fall, accompanied by very disastrous huricanes, which
occurred alternately, or nearly so, in such a manner as to be
only explicable by some great disturbance in the system of trade
winds.   Cyclones swept over the Atlantic, amidst fearful squalls,
whirlwinds and hail-storms.   A tempest also prevailed in France,
Italy and Constantinople.   The winds were of sufficient inten-
sity to detach a stratum of land in places where the surface was
sandy.   This earth carried into the air was certain to be deposited somewhere.   This took place in the south of France.
Ehrenberg, who analyzed samples of this earth, found in them
seventy-three organic forms, some of which were peculiar to
South America, and concluded that its origin was in the new
world.   The interval of time between their leaving America,
Oct. 13th, and their arrival in France, Oct. 17th, was about four
days, which gives a speed of 18¾ yards per second.   The last
remarkable shower of red rain, according to the author above
quoted, was that of Feb. 13th, 1870.

On Feb. 7th, a great barometrical depression occurred in
England.   On the 9th, it had reached the Mediteranean ; on the
10th, Sicily.   This fall of barometer was accompanied by a vio-
lent tempest.   On the 11th and 12th, the weather was calmer
and the barometer reading increased again, the cyclone rag-
ing over the desert of Sahara.   From Africa, the hurricane and
cyclone again made its way back to Europe, accompanied by the
sand swept up from the Sahara.   On the the 13th, a reddish rain
fell at Rome, which, upon investigation, was found to be sand.

The list of cases of this kind might be very much enlarged
upon.   Over twenty of these, with dates and places of occur-

---

*Flammarian, p. 455.

rence, are made mention of by Flammarian. These remarkable cases seem to have occurred in the winter and spring.

Besides the two causes which rationally account for these phenomena—viz., the cyclone or whirlwind, and the traces left by certain kinds of butterflies—a third cause, says the author from whom we have so frequently quoted, must also be noticed ; viz., volcanoes, the ashes of which are sometimes carried to an immense distance. Let us then refer to some of the cases recorded, which are easily attributable to this cause. One of the most remarkable occurrences of this kind was noticed in the parish of Slaine, on the eastern coast of Scotland, January 14th, 1862, a full account of which is given in a little book on Scottish showers, by James Rust. The rain which fell was of two kinds—that of common color, and that of an inky nature, blackening everything as it fell. A remarkable feature of these showers, four of which fell during one day, was the shoal of pumice stones which floated ashore, immediately after the storm. Care was taken to collect the sediment, in places where the chimneys of the houses could not have produced any local effect. Hoffman examined the sediment and the stones, and found them to consist of silica, lime, iron, magnesia, and traces of H_2SO_4 + HCl.

This matter, after a thorough investigation, was proven to be of volcanic origin, and to have come from Vesuvius. The first eruption of Vesuvius, in 79, had its ashes carried to Syria in Asia, and southerly to ·Africa. In 1793 the volcano Skaptaa Jokul, in Iceland, produced a fearful eruption. It covered the island with pumice stones and ashes, and these products were carried by the upper current into Great Britain, Holland, Germany, and in fact all parts of Europe as far as the Alps, were covered with ashes.

Of the active volcanoes of modern times, those of the Sandwich Islands and America have been remarkable. During an eruption of the volcano of Cosagaina, in Guatemala, in 1835, ashes fell upon the Island of Jamaica, 800 miles eastward, and the plains for twenty-five miles were covered with ashes sixteen feet thick ; and ships sailing over 1,200 miles away, were covered by the rain of ashes. The mass of matter sent forth was estimated to be 65,000,000 cubic yards, and the distance covered was estimated to be two million square miles. The Sandwich Island volcano of Mouna Lea furnishes also some remarkable instances of this character, and the ashes, together with what is known as Pelix hair or spun glass, have been carried to immense distances. According to Sir Charles Lyell, from an eruption which took place on the island of Tambawa, lying east of Jamaica, ashes were carried over 1,000 miles, forming a mass two feet thick, through which vessels passed with difficulty. Notices and accounts of the fall of dust traceable to a volcanic origin, throughout Europe, are numerous, and are easily found.

Before concluding the mention of these cases, however, it will be well to note one which is mentioned in the Jour. Chem. Soc., Vol. X.

A cyclone which passed over Africa on the 28th of February, 1872, appeared in Sicily on March 5th. The wind blowing violently on the 9th, 10th and 11th, rain fell, mixed with fine dust. The dried dust consisted of 75 parts argillaceous particles colored by iron, 11.6 parts calcareous matter, and 13.19 parts nitrogenous organic matter. Its specific gravity was 2.52. Examined microscopically, the dust was found to contain an abundance of such organic matter as hairs, fibres, etc., together with five classes of organisms. To the suggestion that the sand might have been derived from the African Sahara, the author replies that his examination of the sand of that desert shows that it has an entirely different composition, no organisms are present, and therefore, in spite of the similarity of specific gravity, and notwithstanding that the latter contains many objects well known in the vicinity, such as the hairs of the olive leaf, etc., he concludes that the volcanic dust was not of African origin, and gives his reason for thinking it to be derived from South America.

We have thus presented, in a very incomplete manner, the facts and circumstances bearing upon the fall of snow colored with dust, in Michigan and the West, and also those relating to events of a similar nature, of known and unknown origin, which are recorded in history. Some of these cases are without doubt attributable to the force of cyclones on the surface of the earth ; others seem to be quite traceable to the force of volcanic erup tions. Under which of these two causes—for it must be accounted for by one of them—the case which we are now considering is to be classed, remains to be decided.

Although at first thought one would perhaps be inclined to attribute its appearance to the first of these two causes, still there is much to be said in favor of either theory ; and although, perhaps, no conclusive theory may be decided upon, a brief mention will be made of those which at once suggest themselves, and the matter be thus left for disposal.

Chemical analysis does not very materially aid in the solu- tion of the problem. The presence of silica, iron, alumina, manganese, etc., neither proves it to be a prairie soil, nor does it prove it to be volcanic. Silica and iron and organic matter form the larger proportion of ingredients; the other elements, which, owing to the small amount of material for analysis, could not be estimated, being present only in traces. The embrown- ing of the substance by heat is due, of course, to the ferric oxide, which by digestion in HCl seems to be all dissolved.

Perhaps the solution of the problem lies in the form and structure of the fibrous and hairy material, which seems to be present in both samples. That some of this is extraneous matter which, either floating in the atmosphere, was taken up by the falling snow, or derived from the atmosphere in which it was evaporated, there can be no doubt. Prof. Harrington, on exam- ining this fibrous material microscopically, was inclined to the opinion that in it were to be detected artificial fibres, probably cotton fibre ; but it is quite possible that this, if present, might originate in the causes above mentioned, or perhaps from the paper in which the powder was wrapped. The appearance of these fibres does not show growth of any nature. Most of them appear to be long, hollow and quite regular tubes, with, of course, a more or less quantity of earthy matter attached to them. In' order to prove the identity of this matter, a comparison was made with a specimen of Peles' hair from the Sandwich Island volcano. This hair or fine spun glass is composed mainly of silica and silicates, and is produced from lava in a state of extreme fluidity, by the force of volcanic action, and the force of the wind, drawing out the melted matter into glass threads and hairs, sometimes smooth and sometimes crisped or curled. This comparison, however, was somewhat unsatisfactory, as far as proving a perfect identity was concerned. Both substances presented very much the same appearance, with the exception of the earthy matter, which was not present in any quantity in the volcanic matter used for comparison. Upon mixing it with earthy matter, the two substances resembled each other very much in their nature, both presenting the same tubular appearance and general appearance. In fact, they were almost identical.

It was suggested to compare the sediment with some of the dust which, blown up from the streets, is deposited on the roofs of buildings. To this end, I collected some of the dust from a clean place on the roof of one of the stores on Main street, and subjected it to microscopic examination. It was made up principally of silica colored with organic and earthy matter, showing plainly the presence of woody fibre, but presenting an appearance quite unlike the sediment in question. Its appearance under the microscope, therefore, according to our eyes, did not prove its identity. It is certainly quite impossible that this substance could have had its origin anywhere in Michigan. It came from some point south-west, and must have been carried across Lake Michigan. The distance across the lake at that point is probably between sixty and one hundred miles. From whatever source it may have originated, then, it doubtless was carried by the upper current, which, according to Dr. Draper, is perpetually blowing over most of the United States from the south-west, at a height ranging above 7,000 feet, and at a velocity which is very variable, but which increases with the height.

On Monday last the dispatches reported the wind to be blowing on the summit of Mount Washington at a velocity of over 150 miles per hour. Assuming that the upper current, on the 5th of February, had a velocity of half this, or 75 miles, it would have taken about one hour for this mass of dust to have been brought from Chicago. Or in other words, if it had been taken up from this neighborhood, this would probably have been done at about 3 P. M. As we have seen from the weather reports quoted on the 5th of February, that the wind all through the West and South-west was not a strong wind, and as it is hardly probable that this great mass of matter could have been carried up by anything but a strong wind, its elevation must have been due to the powerful winds which were noted throughout the West on the 2d and 3d of February, if it was due to this cause.

Being taken up into the upper atmosphere on that date, it may have had its origin in the far south-western or southern part of the United States, or in Mexico, where the surface of the country was not covered with snow.

Reports from this region as to the exact force of the wind we have not been able to obtain. No tornado or cyclone of a destructive nature is mentioned, however, in the papers of succeeding dates. In regard to the possibility of a volcanic origin, the numerous cases recorded of a nature quite as impossible as this, show that there are some grounds for the supposition of such a theory.

The incompleteness of reports which would furnish the exact circumstances existing at various localities at the time of the occurrence, and the very unfinished state of any investigation into the case, would perhaps hardly warrant the adoption of any decided theory. It is, to say the least, a strange phenomenon; and it being the only one to be found recorded as happening in this State, at least, it is not without interest.

9

J.

—— ••• ——

VARIOUS EXTRACTS FROM LETTERS REFERRING TO THE ACTIVITY OF ICELANDIC VOLCANOES ABOUT THE TIME OF THE COLORED SNOW-FALL IN MICHIGAN.

BY W. D. HERDMAN.

[Letter from Jon Olafsson to New York Times(?), written in Iceland, June 16, 1875.]

"Since the middle of December last year, people have noticed many instances of earthquakes every now and then, especially in the eastern and northern part of the island. *About Christmas time*, dense columns of smoke were seen issuing from the mountains in the northeast, just east of the Odadwhraun, in the so-called Dyngjufjöll or Askja, which is situated near the 65° north latitude and about 50′ east of the Ferro-meridian. After the middle of February, an outburst took place about ½° north of the first one. There seems to be a line of volcanic activity from Vatnajoküll, and north as far at least as 65°50′ north latitude, between the Jokulsá and Lake Mijvata. The lava heaps, in some places, are about five miles long and half a mile broad; in other places again, as large as 14 miles long and from 500 to 1,000 fathoms broad. Several visits have been made to the fire, but none by scientific men. Are there no enterprising geologists in the United States who would be interested in visiting the places of the eruptions now, when they seem to have ceased, at least for a time? There has probably never, in historical times, taken place anywhere on our globe

such tremendous eruptions as in Iceland; and this eruption
seems to be one of the most interesting in the experience of all Iceland. The ashes have fallen heavily on eastern districts, in some
places 9 inches think. In about 21,000 square miles it is estimated that 15,360,000 bushels of ashes have fallen. The ashes
from the eruption which took place on the 29th of March, feil
on the following day in Norway and Sweden so thickly that the
sun became black. Several districts, some of the best of Iceland, are destroyed, at least for some years. This will cause
numbers of people to emigrate next year, in all probability.
Fresh intelligence from the fires is expected soon, and as a
steamer will leave here for Scotland in about two weeks, I hope
I shall by that time be able to tell you something more about the
eruption.''

[Letter from Jon Bjarnason to Arthur Macy, written Nov. 27, 1875.]

" Prof. Anderson has told me that you would like to get some
information concerning the last volcanic eruption in Iceland,
this year, as far as known to me. The last letters I have had
from home are dated about Oct. 20th, and Icelandic newspapers
I have got down to the same time. In these last ones there is
nothing spoken of any continuation of the volcanic catastrophes
since the middle of August, when a new crater opened in the
southeast part of the sand deserts of Mijvata (Myvaluorxfi) near
a certain chasm called Sveinagjá. This last eruption was said to
have been very vehement, but how long time it lasted cannot be
seen from the newspapers, and probably people did not know exactly when it ceased, the place being far removed from human
dwellings. But I do not think this eruption has caused any serious destruction, since it has only shortly and incidentally been
mentioned in the papers. How many craters there have been
active in the middle and northeast of Iceland throughout this
year, I cannot tell, nor do I think it is known with certainty to
anybody in Iceland, but in my private letters, which I have had

from various parts of my country during the last summer and fall, it is suggested as a conjecture most probable that the subterranean fire has broken out in 30 to 40 different places.

The most fatal eruption was that which began Easter Monday (March 29), and of which you have certainly read much in English and American papers. Many of the best farming districts in the east of Iceland were partly destroyed by the showers of ashes and cinders, which in this catastrophe were poured out over that part of the country.

You know there is no harvest in Iceland at all, except the hay harvest, and the best part of the hay is raised on small cultivated fields in the near vicinity of the farm houses. Now, if any hay at all could be raised this year in the infested district, the layer of cinders covering the fields 9 to 12 inches deep was to be removed and cleared off. This was done in most places, and as is easily understood, cost the most tedious and hard labor, which indeed in many instances is said to have had its good consequences, but, for a great deal also, was almost useless, because the wind often covered the cleared fields again with as great quantity of cinders as before. The result was that most of the farmers in these districts got some—though very small—hay harvests, and have not been forced to part with all their live stock, as was first anticipated. This year there will not be any considerable need in Iceland as a consequence of this misfortune, both on account of a liberal pecuniary collection gathered on behalf of the damaged people both inside and outside the country, and also because the sheep and cattle which must necessarily have been killed this fall will be sufficient to support them till next summer; but in the following year, it is feared, the bad consequences will be visible, and in the last newspapers there is uttered some anxiety that the hay raised last summer will prove in more or less degree poisoned, and a very unhealthy food for the remaining creatures during the coming winter."

Madison, Wisconsin.   JON BJARNASON.

[From the Penny Magazine, Dec. 21, 1833, p. 196.]

" But the last great eruption (1783) was the most terrific of all that are recorded. This proceeded from the mountain of Skaptaa Jokul * * * At that unhappy season an enormous column of fire cast its glare over the entire island, and was seen from all sides at sea, and at a distance of many leagues. Issuing forth with the fire, an immense quantity of brimstone, sand, pumice stone, and ashes were carried by the wind and strewed over the elevated land. The continual smoke and steam darkened the sun, which in color looked like blood. During the same summer the sun had a similar appearance in Great Britain, and the same obscurity reigned in most parts of our island. Many parts of Holland, Germany and other countries in the north of Europe were visited by brimstone vapors, thick smoke and light gray ashes. Ships sailing between Copenhagen and Norway were covered with brimstone ashes, that stuck to their sails, masts and decks. The whole face of the island has been changed by these terrific convulsions, and Sir Geo. Mackenzie thinks he is safe in estimating that one continued surface of 60,000 square miles has been subjected to the force of subterraneous fire in this part of the world."

K.

ANTIQUITIES OF PERU.*

BY J. B. STEERE, PH. D.

On the old mail route from the Amazon to the Pacific, the tropical jungle reaches, with but little interruption, to the village of Molino Pampa, near the old town of Chachapoyas, where the road breaks out all at once from groves of tree ferns and palms, upon high, cool, grassy plains, with dry wastes intervening, both so characteristic of the Andes.

As soon as the rainy and timbered region is left behind, the character of the remains and ruins left by the ancient peoples who have inhabited the country, changes. Ruins of stone buildings are found, and the dead were buried in tombs of stone and mud, or were put away in caves and crevices of the rocks, instead of being buried in earthen jars, as on the Amazon.

The physical characteristics of a country have much to do with the habits of life of the people inhabiting it, and hence much to do with the remains left by them. People living in regions abounding in stone and with little timber, naturally use the materials most abundant, and leave ruins of huge pyramids and stone temples; while those dwelling in such heavily wooded countries as the Amazon, where stone is rare, build their dwellings of wood, and leave nothing to prove that they have existed

---

* The subject originally treated of was " *The Antiquities met with in a Trip up the Amazon and across the Andes to the Pacific Coast;*" but it seemed too large to be condensed into the space allotted in the printed proceedings of the Association, and that part relating to the remains of ancient races found upon the Amazon has been omitted.

but a few painted earthen pots; though they may have equalled in civilization those who have left much more magnificent proofs of their existence.

While at Chachapoyas, I made a trip with Mr. Arthur Wurtherman, (then in the service of the Peruvian government as provincial engineer,) to visit the ancient ruins of Quillip. We crossed the little river Utcubamba, and rode up the valley on the opposite side. For some distance the water had been carried out over the narrow valley, which was cultivated in corn and sugar-cane; but as we ascended, the valley became narrower and cultivation ceased, while the mountains on each side rose higher and wilder. We now began to see for the first time signs of ancient inhabitants. On the almost inaccessible cliffs, hundreds of feet above us, were tiers of circular and half circular stone walls, apparently from ten to twenty feet in diameter, and four or five feet in height; probably the foundation walls of houses, the roofs and superstructures of which had been made of grass and wood. The first impression one gets, is of the warlike nature of the people who inhabited the country in those days, and the continual state of fear in which they lived. Their fields were in the little valley below, and they must have spent hours every day in climbing to these aeries in the rocks. Peru is now one of the most unsettled and revolutionary states of the earth, but its towns and villages are now built in the plains, without walls to defend them.

The immense population that must have existed here was shown by the frequent tombs. These were built in the shelter of the cliffs, or in little depressions in their sides, where stones and mud had been carried to build up little niches, into which the bodies of the dead had been crowded. Nearly all of these had been broken open in the search after valuables, and scattered bones and bits of the cotton wrappings of the dead were all that remained. The caves, that were frequent in the limestone rocks, were also filled with human bones.

In a perpendicular cliff on the other side of the river in one place we could see a number of holes like the mouths of mines, and we were told that these also were tombs. The impression is forced upon one, as he finds such immense quantities of human remains, as well as signs of cultivation, in places where water must have been carefully brought for many miles for irrigation, that the human race must have at one time filled up and overrun the territory of Peru, especially the country near the coast, as it does now in China, making it necessary to use every possible means for the support of life.

The country became higher and rougher as we proceeded, until we turned off up a little branch of the river, running down between immense wall-like cliffs of limestone, that seem to have been rent apart to give it a passage. The strata in these walls of rock were most curiously distorted. The great bends extending for miles, the same layers of rock at the higher parts being several hundred feet higher than in the lower. We had great difficulty in scaling the steep sides of these mountains, but finally found ourselves in the higher, cooler regions above, where the surface of the country, though still rough and steep, was much moister than the valleys below.

I had wondered, when below, why any people had built such a great fortress as Quillip was said to be, in these almost inaccessible mountains, that appear from below to be nothing but barren rock; but I found there were great extents of country up here, covered with a rich dark soil, which was in some places cultivated in wheat and barley, the rougher places growing up to thorny bushes that no where reached the stature of trees. Indians were plowing as we passed, with cattle yoked by the horns to rude wooden plows with one handle. It was only after repeated scratchings with these rude implements that the surface of the fields took on anything like a cultivated look.

These mountains, wherever of any value, are owned in large estates by the descendants of the Spanish conquerors, who keep a lot of Indians at work cultivating the lands and tending cattle, while the proprietors live in Chachapoyas, or, if able, in Lima.

The second day of our journey we reached the estate upon which the fortress is situated, though it was at a distance of several miles of rough mountain roads from the house where we stopped. This, a rough stone building covered with grass, was the place where the proprietor stopped when visiting his estate, and was occupied by the overseer, who seemed to be a poor relation of the proprietor. After a night's rest, broken somewhat by the attacks of garapatos (a peculiar species of ticks, the bite of which produces inflamed and painful wounds), we set out under the guidance of the overseer and several Indians, who followed on foot, to visit the fortress. We had to cross the ravine by the most dangerous roads, where the most cautious of us dismounted and climbed on foot. We passed a low, dirty village of stones and mud, covered with grass, where the Indians of the estate lived, and after an hour's riding through thick bushes, came out in sight of the fortress, a long, low wall surrounding the crown or ridge of the mountain. As we rode near it we estimated it at half a mile in length and perhaps a quarter of a mile in width, with walls from thirty to sixty feet in height. On the side on which we approached there was a peculiar gateway in the wall. This was some six or eight feet in width at the bottom, but gradually growing narrower, until at a height of twenty feet the walls nearly touched, and had probably originally done so, forming a pointed arch, if it could be called such. The wall was of regular layers of limestone slabs, laid up without mortar. The stones were from one to two feet in thickness and two to four feet in length. They had all been carefully worked, but apparently bruised into shape with some dull, blunt instrument, rather than cut, the corners being all somewhat rounded, though the joints were close. As we entered the gateway we found ourselves in a walled passage, open above. We gradually ascended, the passage growing wider and then suddenly narrower until, at the top, but one person could pass through at a time. Passing out here, we found ourselves on a plateau, stone and earth having been carried up and filled in, until the whole interior of the fortress was built up to the height of the walls. Another gateway and passage led up from the opposite side of the fortress, and opened above within a few feet of where the one we had ascended came out, making it possible for a few armed men, standing here, to defend the whole fortress, for these seem to have been the only ways of approach.

There were great quantities of thorny bushes growing over the top, but among these were many ruins of little round houses, with their walls of mud and broken stone laid up to a height of four or five feet, where there was generally an attempt at ornamentation, the top stones being arranged in patterns and figures. What the roofs of these dwellings had been we could only conjecture from the grass roofs of the inhabited villages around. Turning toward the west end of the fortress, we found a curious round tower, made of the same hewed stone as the principal walls, but larger at the top than at the foundation. It was thirty feet in diameter, and about twenty feet high. The walls of the gateway, where remaining in place, were perpendicular, and at the base there was a rude human face carved in a stone of the wall, on each side of the entrance, (the only signs of sculpture we saw.) In digging about the base of this tower we found quantities of broken human bones and bits of painted pottery, and a foot and a half below, a pavement of stone slabs. Along the edge of the fortress on this side were other walls, apparently of large buildings, and in these were many openings, into which human remains had been crowded, the bodies having been doubled up so that the knees touched the breast ; and in this way five or six bodies had been crowded into a space where one person would have been troubled to sit comfortably. The cotton wrappings of the dead, and in some cases the hair and shrunken flesh, still remained.

Turning toward the east, and passing the place where the two entrances opened upon the top of the fortress, we came to the foot of a second great wall, like the first, from thirty to sixty feet in height, and made like the first of large cut stone laid up in regular layers. It covered about a third part of the space of the main fortress, and had been filled in in the same way, level to its top, with earth, forming a second fortress mounted on top of the first. As we scrambled along the foot of this wall, we found it had been used as a place of interment, and apparently by a later race than the original builders, the large stones of the wall having been pried out of place, and the bodies of the dead put into cavities behind, the original stones being in some places put back, in others smaller stones and mud being used. One of the Peruvians with us ventured the opinion that it was done by the original builders, to frighten their enemies when they should try to tear down the walls, and should find them filled with human bodies. This second fortress probably had much such an entrance as the first, but the wall had fallen where it had stood; and we climbed up over the fallen stones, to the top. Here we found again numbers of foundations of round houses, and low walls filled with niches containing human remains. There was no wall here on the north side, for some distance, there being a perpendicular cliff that defended this side. We followed along this until we came to the north wall, which was broken down where we reached it, and descending here we followed around to the gateway again, where we had left our horses. Mr. Wurtherman estimated that it must have taken twenty thousand men at least twenty years to build this fortress; but with the rude means they must have had for cutting stone and transporting earth, it probably took much longer.

The next morning we visited another side of the same mountain, where there were other walls and ruins. After an hour's ride we found ourselves at the foot of the mountain. For three or four hundred feet it rose as steep as is possible to climb, but having a few rocks and bushes to which we could cling; and then there was a perpendicular cliff of several hundred feet, and on projecting ledges of this were two walls of cut stone,—one quadrangular, and the other crescent-shaped. They were eight or ten feet in height, and had openings at regular intervals, that looked very much like loop-holes. After a hard scramble we reached the base of the cliff, and found ourselves some sixty or eighty feet below the walls, and no apparent way of reaching them. We followed the base of the cliff for quite a distance, finding caves filled with human remains; and finally, by climbing a small tree that grew against the side of the cliff, reached a narrow ledge that led up toward the walls. This was so narrow that in some places we had to creep around projecting points of the rock, but we finally reached the foot of the lower and quadrangular wall. It was made of cut limestone slabs, like

those of the great fortress of Quillip, but smaller, and laid up in
mud. A ledge of the cliff, not more than four feet in width, had
been used as a foundation for this, the wall being built on the
outside of this, and leaving barely room behind for a person to
pass between the wall and the cliff. It had been built up in this
way some eight or ten feet, to where it reached the level of
another ledge which gave room for a superstructure, which was
probably of wood or grass. The narrow foundation wall was
supported by small pieces of wood, which were built into it, and
ran back into the rock behind, where holes had been made for
them. These pieces of wood were still sound, though they must
have been there for several centuries, or at least from before the
Spanish conquest.

The appearance of loop-holes was made by stones being
drawn back from the general face of the wall, there being in
reality no openings through it. The crescent-shaped wall was on
a ledge still above ; and some fifteen or twenty feet over it, and
sticking out horizontally from the cliff, was a bar, that has a great
celebrity in the country about, as it is supposed to be of gold,
and enchanted, etc. As near as I could make it out, it was
nothing more than a wooden bar, that had been used to support
the building that must have originally stood upon the wall, it
having been so well secured in the cliff that when the building
fell it had remained as a continual wonder for the simple Indians
of the country.

These ancient ruins along the Utcubamba and at Quillip
are east of the Amazon, and are separated from the ruins about
Cajamarca and the coast, by the central range of the Andes and
by the river itself, which is swift and dangerous, and now only
crossed by means of large rafts. Whether they are to be classed
as the work of the Quichuas or other coast or Andean peoples,
or have been made by nations coming from the East, who changed
their mode of life when they reached these high, cool, rocky
regions, is a question that is still to be decided. I have seen no
account of Quillip being inhabited at the time of the Spanish
conquest, and it was probably a ruin then.

The trip to Cajamarca, from Chachapoyas, led up the
Utcubamba, but the valley was most of the time too rough and

narrow for cultivation. As we reached the foot of the pass of Calle Calle, we found ruins of large buildings, built roughly of boulders. They were probably intended to guard the pass.

Cajamarca was one of the principal cities of the Incas, and the place where Atahuallpa, the last of the line, was put to death by Pizarro; but there are but few remains of its ancient inhabitants. I was shown an ancient building which was said to be the identical room in which Atahuallpa was confined, and which he offered to fill with gold vessels as his ransom.

The building is perhaps twenty-five feet long by sixteen wide, and the old wall is about ten feet high; but it has been built up with adobes and covered with tiles, and serves for a modern residence. The walls were not made perpendicular, but were drawn in a little on every side, giving it the appearance of a truncated pyramid. The stones of which it is constructed are not squared, but have an irregular number of sides, giving the walls a curious appearance. They were laid up without mortar, but the joints are very close, and it has been supposed that they were rubbed together until they were fitted. There are many articles of pottery and human remains dug up about Cajamarca, but the valley is so moist that such things are not so well preserved as in the coast country.

The greatest amount of ancient ruins found in Peru is in the rainless district between the last range of the Andes and the coast. This is made up of the ancient bed of the sea, as about Trujillo, and of the valleys of streams that come down from the mountains behind, and varies in width from one or two to thirty miles. It is now most of it desert, the sand blowing over much of it in whirling heaps. But there is every reason for believing that it was anciently all under cultivation, and most thickly inhabited. Excavations in the sand at almost any point uncover great numbers of human bodies, preserved by the dryness of the climate, like Egyptian mummies, with pots full of beans and corn and peanuts, and other articles of food. Besides these buried remains there are immense numbers of ruins of adobe or sun-dried bricks, these in some cases extending for miles. They have stood here for centuries, with no rain to wash them down, rounded a little by the winds and moving sand, but under the circumstances almost as imperishable as granite. In some cases these seem to have been walls for defense, in others they are great pyramids, while in others they are remains of temples and dwellings. Little or no stone ruins are found in this coast country, though there is plenty of granite near, and the reason seems to be that these sun-dried bricks were found to serve as good a purpose, while they were so much easier made. The ease with which these buildings were made, and also the ease with which they could be destroyed by an attacking force, led to their being made very thick and massive; and there are great mounds of them in the valley between Callao and Lima, that are generally supposed by travelers to be natural hills, and artillery is said to have been posted upon some of them in certain of the revolutions the country has passed through.

There seems to be but little doubt that this whole district was once under cultivation, as bits of pottery and signs of occupation are found all over it, while the steep valleys of the streams farther back in the mountains have been terraced up with great labor, to save a few acres of ground for cultivation; but under the present system nine-tenths or more of the level lands along the coast are desert. Even much of the great level valley about Lima is waste, and this is said to be from lack of water. Whether this lack of water arises from a real change in the rain-fall in the mountains behind, and a lessening of the rivers in this way, or whether the trouble is in the present careless method of distribution of the water, is a matter of doubt, though there is some reason for thinking that at some time there was a much greater amount of rain in the mountains, and that the rainy belt approached much nearer the coast. In many places there are torrent beds in the mountains, where no water now flows at any season. Much of the water is certainly now wasted, from the irregular way in which the farms are laid out and the carelessness shown in the use of the water.

The Incas or Quichuas are generally credited with all of these ruins along the coast; but it seems probable that several powerful races have occupied the coast country in different places and at different times, and that the Incas were inhabitants of the cooler plateaus of the interior. Their principal cities of Quito, Cajamarca, and Cuzco, were all on these plateaus, and the remainder of the race are now found inhabiting the cold mountain valleys, or upon the estates of the interior, employed as peons or serfs. The whole Spanish account of the conquest, and of the people and their customs, numbers, cities, etc., seems a bundle of lies, conflicting among themselves, and not at all corresponding with the real facts, where they can now be investigated. This leaves the question of the origin and distribution of races in South America open to the theories and speculations of every one; and the graves and ruins give, and will continue to give for a long time to come, vast quantities of material to found theories upon.

The graves that I examined seem capable of being divided into two classes; but whether the difference was caused by difference of race or difference of locality and surroundings, is doubtful. The general style of burial on the plains was in vaults beneath the surface. In finding them, one generally digs through two or three feet of sand, sometimes finding bodies lying prone in this, wrapped in cotton cloth; and then coarse mats of rushes or bamboo are reached, supported upon poles, that lie across the vault below. In one end of this vault the bodies are found, appearing to be bundles of cloth, standing on end, and these bundles are often enveloped in baskets or sacks of coarsely plaited rushes. Upon unwrapping these bundles the bodies are found to have been doubled up so that the knees nearly touch the breast. In the graves at Pachacamac, south of Lima, there were found sticking in the top of these bundles figures to imitate the human head, the face being carved rudely in wood, with shell eyes, and a shock of fibre of some kind being used for hair. At the feet of these bundles were generally found a number of pots, often six or eight, or more, covered with dishes made from gourds or squashes, and filled with peanuts, beans, Indian corn, roots of cassava, and in one case, of the skeletons of a small animal, probably a guinea pig. Cotton is also often found. The pots found, filled with food, have been used in many cases,

and are blackened with fire. There are also found plates of
gourd, and rudes ones of pottery.

The finer kinds of pottery are much rarer, and the only
specimens I succeeded in finding were wrapped up with the body.
This fine black pottery—for it is nearly all of this color—seems
to have been in nearly all cases used for water jugs or bottles, to
be carried about at the girdle ; and I have seen one of these that
represented a human figure with one of these miniature jugs
hanging from the belt. For this purpose they are all provided
with handles, or double necks, for being carried. There is a
wonderful variety of them, a collection of hundreds of them
having hardly two alike, while they are made to represent all
kinds of animals, fish, turtles, seals, serpents, birds, monkeys,
men, fruits, roots, etc., etc.

In spite of this variety, there are several general forms that
are often repeated. One of them is a bottle with a double
throat, that unites above to form the mouth. Another is the
double jug united by one or two hollow tubes ; these are often
in the form of birds or animals, and often have whistles arranged
in the interior so that in drinking from them they whistle.
Another form is pointed at the bottom, so that it will not stand
upright unless stuck into the sand—imitating in this some of the
ancient pottery of the East. Little earthen images five or six
inches in height are often found. These are all alike in form,
and in having a peculiar head-dress, and ears enlarged, as they
are now found among some of the wild tribes of the interior of
Peru. These can have served no other purpose than that of idols.

The pots, besides being made, many of them, in the shape
of men and beasts, are nearly all ornamented with painted or
raised figures. Some of them are covered with little figures in
relief, of fish and birds, the same figures being repeated many
times. The strange, chair-shaped marking still used by the
natives of the Amazon, and found on Chinese and Egyptian
pottery, is also frequent. This black pottery is quite porous, and
is often used by modern Peruvians for keeping water cool. Some
of these pots, at least, seem to have been made in halves, and
then stuck together.

The art of working it has been completely lost; and though imitations are made at the present time, and colored black, they are very easily known from the ancient ones. The style of burial and of pottery described is found along the coast at or near Lima, at Pachacamac, and north at Trujillo and Pacasmayo.

In the mountains behind, where the country was much rougher and stone abounded, I found no pottery buried with the dead, though this may be by no means general. Many of the dead are buried in little niches built up in the rocks, of stone and mud. Wherever there is a projecting rock, it is taken advantage of, and hollows are dug out beneath it, where the dead are stowed away. Up the Rimac River, above Lima, at Chosica, and still further up, where the river valley narrows to a mile or so in width, the mountain sides behind have been terraced up for hundreds of feet with strong stone walls five and six feet in height, while there are remains of ditches along the mountain side, where the water has been brought from long distances above, to irrigate these terraces. Besides the remains of burial places in the rocks, in this locality, the ancient inhabitants seem to have buried in and beneath their own houses. The village of Chosica was built on a steep mountain side that was thickly covered with boulders, some of them of great size. Among these they built a town that must have been much like a great ants' nest. The houses were very small, many of the rooms being six or seven feet long by four or five wide—little cells, burrowed out from the rocks, or built up out of them. They seem to have passed about the town along the walls of the houses, which were probably, from the remains, covered with corn stalks. They burrowed out vaults beneath these houses, for the dead, and in some cases filled them into the lower rooms. These lower vaults were usually walled with boulders, and were covered with flat stones, that were bound together and supported by other stones piled upon them on the outside, while a large central flat stone seems to have covered the main hole left for entrance. There were also, in most or all cases, little holes a foot or more square, that opened from the side into other vaults near.

In these tombs the dead were found bundled up, as in the others, and, standing up in one end, often to the number of fifteen or twenty. The bodies had been doubled up, as on the plains, and wrapped in many folds of cotton, some of this striped and colored. It seemed that they had been in the habit of opening these tombs and re-wrapping the bodies of their friends in new cloths, while the faces had often been painted red, as if the living had tried to make them look better. They were buried, men, women, and children, indiscriminately, each with their peculiar arms or implements. The women usually had boxes with balls of yarn, and needles of the spines of agave, or of bronze, with bits of half-woven or netted cloth; while the men had slings for slinging stones, about their necks or in their hands, while we found one or two with dents in their skulls, evidently made by stones slung, that caused their death. The men also generally wore about the loins a leather dress, with a pocket, in which was usually found a little gourd of quicklime, for using with the coca, while a cotton bag at the side contained a supply of coca leaves. The little boys had small slings about their necks, while the girls had miniature needles and work-baskets. One little child, buried in its mother's arms, was wrapped entirely in cotton, and there were many other proofs of the care with which the dead were treated in those days.

The most of these vaults, as well as those in the plains below, have been broken into by the Spanish and present Peruvians, in the search after valuables. They call this old people " infieles " —heathen—and do not consider their graves worthy of respect, though they are descended from them in part; and they heap out these remains by the hundreds, and allow them to whiten in the sun and wind.

There is a place near Lima where one passes for half a mile through arms and legs and trunks, and grinning heads covered with hair, that have been heaped out in this way; one of the most horrid sights one can imagine. Many of the heads of these people have been pressed out of shape, ordinarily, as it seems, by pressing upon the forehead and the base of the skull behind, giving the head a wedge shape. In many cases the brain was crowded out over one or the other ear, giving the skull a curious lop-sided appearance. This custom does not seem to have been universal, as normal skulls were found in the same villages and in the same graves, with the ones that had been pressed out of shape.

While making a journey over the Andes along the route of the Lima and Oroga Railroad, I saw a method of cultivation of the soil that is probably the one anciently used by the Incas, before the introduction of horses and cattle by the Spanish. They then possessed no beast of burden but the llama, and this animal is too small and weak to have been used in the cultivation of the ground. When we had reached a point where the valley of the Rimac River is too narrow and too cold to be any longer an object of desire for the Spaniards and their descendants, we found several Indian villages, the inhabitants living from their flocks of sheep and llamas, and from little patches of potatoes and quinoa. The hill-sides were terraced up, as below, probably the work of centuries past ; and one of these terraces, too narrow to plough, two Indians were cultivating with implements that looked like plow handles, being curved above, to take hold of, and shod at the lower end with iron, and with a support for the foot, tied on with thongs. They raised these narrow, spade-like implements at the same moment, and stepping each a step to the right with military precision, they set the spades to the ground and threw their weight upon them, driving them in six or eight inches, and then at the same moment pried back upon them, loosening quite a sod ; and then they stepped again to the right, loosening at each time about as much earth as one would with an ordinary spade. An Indian woman followed, on her knees, and turned the sods bottom side up with her hands ; and I concluded that I was looking upon the identical method of cultivation of the ancient Peruvians.

L.

THE HISTORY OF THE DOCTRINE OF SPONTANEOUS GENERATION.

BY EDWARD S. DUNSTER, M. D.

In connection with the very able paper of Dr. Lionel Beale, on the nature of life, read at our last meeting, it has occurred to me that a historical sketch of the rise, progress, and present status of the theory of spontaneous generation might be of value. We cannot approach the study of the wonderful mystery we call life without coming, at the very outset, face to face, with the problem of its spontaneous origin; and we must examine and either set aside or accept it, before we can make headway with the higher questions involved in such study. It is instructive, also, for us as students of science to occasionally survey the past, and observe the slow approaches by which our present knowledge has been attained. It gives us an insight into the character of our work, and compels a higher appreciation of its positiveness, when we see that it has been gathered literally by centuries of patient and cautious investigation, in the process of which error after error has been eliminated; and thus, steadily though very slowly, there is a nearer approach to ultimate truth. Such a retrospect may well serve to restrain the impatience of those who are disposed to scoff at science by reason of its changing phases, for it is the distinguishing characteristic of true science that she does not let a belief or theory encumber her progress when fuller investigation has shown that such belief or theory is no longer tenable, but she sweeps it away as remorselessly as the whirlwind crushes down the forest in its advancing track, and rejoices that "the grave of each superstition which it lays is the womb of a better birth." *

I do not purpose, however, to enter into a discussion of the arguments for or against the doctrine of spontaneous generation. Such a task would require a series of lectures, instead of the limited time allotted to me this evening, and it is doubtful, too, if the resulting gain would be at all commensurate with the labor, for it is a question which in the nature of things cannot be argued on the grounds of authority or of probability, but must rest on experimental evidence alone. In the interpretation of this evi‐ dence, however, we may with propriety accept the opinions of those who by long training in scientific research are best quali‐ fied to estimate such evidence at its real value.

The history of the doctrine of spontaneous generation may be conveniently divided into three epochs. The 1st covers the period from Aristotle, 325 B. C., to Redi, A. D. 1668. During this epoch spontaneous generation was believed by all natural‐ ists to be the common mode of the production of a very large class of animals. The 2d epoch extends from the time of Redi to the experiments of Schwann and Schultze, in 1836–7. This epoch presents two phases, one relating to the generation of animals visible to the naked eye, the other relating to the gener‐ ation of infusorial animalcules invisible to the unaided eye. As regards the first, spontaneous generation during this epoch " was narrowed down to a rare and exceptional mode of the reproduc‐ tion of a few only of the most obscure species, and finally shown to be untenable even for them." † The generation of a large share of the entozoa was also explained during this period, and they were removed from the class formerly believed to be pro‐ duced by spontaneous generation. As regards the other phase of this epoch, that relating to the infusoria, it may be said that

---

* MAUDSLEY: *Body and Mind*; London, 1870; p. 111.

† J. C. DALTON, M. D.: *Spontaneous Generation, New York Medical Journal*, February, 1872. I am indebted to this admirable paper of Prof. Dalton's for much of the material here made use of, and I desire now to express my acknowledgments for such use in the instances where I have not given the reference and page.

the mode of their generation was quite fully explained, at least for all but the very lowest species, and that generally scientific men held that the question was put to rest by the decisive experiments of Schultze and Schwann, just alluded to. The 3d epoch dates from the year 1858, when the question was reopened in Paris for special reasons connected with the study and theory of evolution. In this, the present epoch, the whole battle ground is within the domain of infusorial life, and although, by the majority of scientists, the victory thus far is conceded to the advocates of biogenesis, or life from preëxisting life, a few still courageously contend for the opposing theory—abiogenesis—life without preëxisting life, or life from inorganic matter alone.

1st Epoch. During this period a belief in the spontaneous generation of many animals was universal. Even Aristotle, who may be considered to represent the highest type of the scientific culture of the day, divided animals, with reference to their mode of production, into two classes. The one was derived by succession from preëxisting parents, life being transmitted either by the production of living young resembling the parents, or through the hatching of eggs, or, as in many insects, by grubs or larvæ. In the other class no such connection could be traced, and hence they were considered to originate spontaneously from "the fortuitous concourse" of inorganic materials, from the slime or ooze at the bottom of the sea, or from the decomposing remains of other animals. Thus the shell-fish, such as clams and oysters ; the sea-nettles and sponges ; the maggots that invariably swarm after a time in dead meat ; very many of the smaller insects that appear so suddenly ; grubs, moths, eels and many other small fishes, are enumerated as originating in this way. The idea of decomposition and recomposition of organic atoms was a favorite one, not only during antiquity, but down through the middle ages. It finds its best expression, perhaps, in Aristotle's well known formula, *Corruptio unius est generatio alterius.* These crude beliefs were not confined to the Greek scientists alone, but continued even down to the middle of the seventeenth century, at which late day Kircher, the learned Jesuit, declared that to produce a crop of serpents it is only necessary to pulverize one and sow the powder as seed in the earth. He further averred that fragments of plants falling into water became transformed into animals, and he actually figured such animals in his book.* Van Helmont, too, we find describing a mode for the artificial propagation of mice, frogs, and eels.

With our present knowledge of the mode of reproduction in animals, we may perhaps smile at these crude and incorrect notions, but we must remember that they were the best conclusions then attainable, and they were the result of a truly scientific but imperfect study of natural phenomena. Dr. Dalton well says: "Aristotle represented in natural science, as in so many other departments, the entire scope and successful activity of the Grecian intellect. He occupied the position which was afterward held by the Buffons, Linnæus, and the Cuviers of more modern periods; and it is certain that the opinions which he expressed must have seemed reasonable from his point of view."

It is not out of place to mention here some of the causes of error which are now apparent. The young of many of the lower varieties of animals are so wholly unlike the parent, that it was impossible to trace any similarity or relation between them, until after patient observation the intermediate stages in their development were learned. A familiar illustration of this is the larval form of the common butterflies and moths, and the varied appearances seen in alternate generation in insects. In their successive developmental stages the animals will often inhabit different localities, and in some instances even different elements. The secretive habits of many of the oviparous and viviparous animals precluded for a long time knowledge of their mode of reproduction. Some, as for instance fishes, will migrate long distances, deposit their eggs quickly, and as suddenly disappear. After a time the ova are hatched by the favoring influences of light and heat, no parent animals being present in the vicinity. In others the young on being hatched quickly betake themselves to a different locality. Again, ova not infrequently lie dormant, as it were,

---

* *Edinburgh Review*, Vol. 89, p. 167.

for a season, or even for a period of years, and when finally developed the parent animals have long since disappeared from the
face of the earth, and hence an easy belief in a new or spontaneous generation.* Gradually, however, after years and even
centuries of patient investigation, all these and the kindred difficulties have been removed and the errors have been explained, so
that even during this first epoch some of the supposed cases of
spontaneous generation were removed from this category, and
explained in accordance with the increased light thus gained.
This brief survey of the first epoch in the history of spontaneous
generation must suffice. Indeed, thus much of reference to it is
only pardonable, for the purpose of contrasting the opinions
then prevailing among scientific men, with the more positive
knowledge which now obtains.

2d Epoch. The first solid and experimental advance toward
the positive knowledge of to-day, and the first distinct repudia-
tion of the doctrine of spontaneous generation, was made in
1668, by Francesco Redi, the Italian naturalist.† "He did not
trouble himself," says Huxley,** "with speculative considerations, but attacked experimentally what had been considered to
be particular cases of spontaneous generation." He directed his
attention first to studying the origin of maggots in putrefying
meat. He observed that before the appearance of such maggots
flies were invariably to be seen hovering about and alighting

---

* " A remarkable instance of this is the case of the American seventeen-
year locust (*Cicada septendecim*), where a period of seventeen years elapses
between the hatching of the larva and the appearance of the perfect insect;
the larva all this time remaining buried in the ground, while the life of the
insect in its perfect state does not last over six weeks. A brood of these
locusts appeared in the city of New York and its immediate vicinity in 1843,
and again in 1860. If they return with their accustomed regularity, their
next appearance will be in 1877."—DALTON: *loc. cit.*

† FRANCESCO: An Italian physician, 1626-1697, distinguished alike for his
attainments in literature and in natural history. His writings have been
collected and published in a single volume. *Opuscoli di Storia Naturale.*
Florence, 1858.

** *President's Address to British Association*, at Liverpool, Sept., 1870. This
address has been published in many journals, and also in separate form.
Nature, Sept. 15, 1870, p. 400.

upon the meat, and he suspected that they were the progenitors of the maggots. In midsummer he took a number of wide-mouthed jars, and placed in them bits of flesh. Some of the jars were left open—some were covered with paper carefully secured around the neck. Maggots soon appeared in the open jars, but none were seen in the closed jars, even after weeks had elapsed, while the flesh continued to putrefy just as in the other set. Then using fine gauze as a covering for the jars, the result was the same. His mode of argument, therefore, was, that the cause of the formation of the maggots must be something that is kept away from the meat by the gauze. This something must be solid particles too big to go through the gauze, for air and fluids will readily pass through. Nor can there be any doubt as to what this something is, " for the blow-flies attracted by the odor of the meat, swarm around the vessel, and urged by a powerful, but in this case a misleading instinct, lay eggs out of which maggots are immediately hatched upon the gauze." *

These experiments were repeated with a great variety of substances, and with various modifications, but the results were uniformly the same, and so far as they went they carried conviction ; but it must be remembered that they disproved spontaneous generation only for the special cases under consideration. The presumption, however, that all instances of the supposed origin of life from dead or inorganic matter might be in a similar manner explained, by the introduction in some way of living germs, rapidly gained ground, and was enunciated by Redi himself as at least probable. He even suggested that in this way we might explain the generation of the entozoa, or internal parasites of animal bodies. Redi was followed by Swammerdam** and Vallisnieri,*** who repeated his experiments, and the combined

---

* HUXLEY: *ibidem*.

** JOHANNES: a Dutch physician and entomologist, 1637-1681. He was one of the earliest to make dissections of the human body. He published a number of entomological works. His "History of Insects" was claimed by Boerhaave, his editor, to be incomparably superior to anything that had preceded it. An English translation was published in 1758.

*** ANTONIO: an Italian physician and naturalist, 1661-1730. He studied medicine under Malpighi, and was subsequently Professor in the University

result of their writings was to entirely subvert the belief in the spontaneous generation of insects and all animals of a higher organization. Since their day no one of any scientific pretensions has ventured to propound this theory for any species of animal life with a high grade of organization.*

The discovery of the mammalian egg which dates from 1673, by De Graaf, of Delft, in Holland, and the full history of its mode of development which was closed by Von Baer, in 1827, threw a flood of light upon the general question of generation, and stripped it of the mystery which hitherto had been carelessly supposed to surround it in the highest orders of animals, especially in man himself. Its bearing, too, upon our subject is obvious. In this way " spontaneous generation lost its rank as a great natural division of the reproductive function ; and came to be regarded as an exceptional phenomenon, confined to a very few species whose existence could not be accounted for in the ordinary way. Its territory was narrowed exactly in proportion as the knowledge of natural history advanced ; and it became reduced almost exclusively to the class of animals known as *entozoa* or internal parasites."**

These are organisms, some of them microscopic in size, that live within and prey upon the bodies of other animals. They are found in special habitats or organs, and each species of animal has its own particular parasite. Thus, confining our illustrations to a few only of those met with in the human body, we may notice the different varieties of solid and hollow worms (*sterelmintha* and *cœlelmintha*) that infest different portions of the alimentary canal ; the *trichina spiralis* that dwells in muscle ;

---

at Padua, and was especially celebrated for his researches into the various systems of generation. His works were published in three volumes folio, at Venice, 1733.

* Mr. Crosse's electrical spiders (*Sequel to Vestiges of Creation*), a kind of " microscopic porcupine," which he asserts were developed in a solution of silicate of potash, through which the constant galvanic current was continuously passed for two years, are unworthy the dignity of a serious refutation. A writer in the *Edinburgh Magazine*, April, 1867, humorously depicts this, and very appropriately too, as a most singular case of delusion.

** Dalton: *loco citato*, p. 121

the *strongylus gigas* that makes its abode only in the hilum of the kidney; then there are others peculiar to the brain, the liver, cellular tissue, etc. These creatures long puzzled and completely defied the naturalists in their efforts to explain the mode of their origin, and it is a curious study now to look at the shifting opinions which from time to time have been entertained regarding them. To illustrate: Linnæus, the celebrated naturalist, thought that the internal parasites were terrestrial or aquatic animals that had been swallowed with the food or drink. Bremser and Rudolphi, after twelve years of research, disproved this by showing that there was nothing in common in organization between such parasites and any known species. Boërhaave suggested that there was some metamorphosis or monstrous growth that occurred in them in their new and unaccustomed habitats. This was a leaning toward the truth, for we do find remarkable changes in successive stages of development, but the error was in the starting point.

Without dwelling further upon their opinions or without an attempt to detail the progress of the study, it is sufficient for my purpose to say that at last all these parasites were found to come from eggs, and in turn to produce young by sexual generation.* Years upon years of the closest investigation were necessary to complete this study, and the nature of the difficulties to be contended with were such that it seemed almost impossible to overcome them. This is well illustrated by the cysticerci, the intermediate stage or larval forms in the development of the tapeworms. They live in a closed cyst in the solid tissues, and they are absolutely sexless and unprovided with generative apparatus. To connect them, then, with the mature parasite, which lives in the alimentary canal alone, was a difficult task. The painstaking

---

* As late as 1858, Pouchet, the uncompromising advocate of the theory of spontaneous generation, questioned the truth of these discoveries in the generation of parasites. Says the writer in the *Edinburgh Review* (*loc. cit.*), " like a true Frenchman of the feebler sort he says, " *tant pis pour les faits!*" and rejects the facts which reject his hypothesis. He doubts the truth of these discoveries, " the monopoly of which." he naively says, " has by a singular anomaly belonged to foreigners." This reminds one of the pious patriotism of Lamartine, who said that when God has a noble idea to vouchsafe to mankind He always puts it first into the brain of a Frenchman.

labors of the helminthologists finally determined the mode of their origin, and completed the record of their natural history, by showing that, for the full round of their development two animals are necessary. The second of these usually stands to the first in the relation of prey or food. The mature parasite lays eggs in the alimentary canal of the first animal. These ova are swept out with the alvine discharges, and through the medium of surface water or herbage, some of them find their way into the alimentary canal of the second animal. Here the ova find conditions favorable to the first stage of their develop- ment, and they are now provided with a boring apparatus by which they make their way through the walls of the canal, and travel long distances, finally to ensconce themselves in the solid tissues where they become encysted. The second animal being killed, its flesh is eaten by the first. The cyst wall is digested, and the cysticercus, thus freed from its environment, now finds the appropriate nidus for the final stage in its development. Thus each tænia has its own cysticercus whose distinctive characteristics can be recognized under the microscope, and furthermore the tænia, peculiar to one species of animals, is never found infesting any other species.*

One can never sufficiently admire the splendid patience of such men as Diezing, Kuchenmeister, Haubner, Von Siebold, Leuckart, Van Beneden and others, who almost literally devoted their lives to these studies. The details of their experiments, both on man and on the lower animals, and their cautious, long- continued, and at times unpromising researches, form one of the most entertaining as well as instructive chapters in the whole re- cord of natural history study.† These labors, it is true, were

---

* Van Beneden's little book, *Animal Parasites and Messmates*, published since this lecture was given, furnishes for the English reader an excellent ac- count of the development and migrations of these entozoa. New York: D. Appleton & Co., 1876.

† An amusing illustration of the precision of the results obtained by these investigators is found in the well-known narrative of Van Beneden* in his monograph upon Intestinal Worms. For the purpose of illustrating the

* Van Beneden: *Mémoire sur les Vers Intestinaux*, Paris, 1858, p. 155, and *Animal Parasites and Messmates*, pp. 71 and 222.

not completed during the epoch under consideration, but during it, the presumption was clearly established that on fuller investigation a solution would be found of all cases that had hitherto baffled detection. The latest of these investigations worthy of note are those of Leuckart (1856–7) and Virchow (1858) upon the *trichina spiralis ;* and they have confirmed in a very positive manner the opinion just stated, for it is a reasonable and almost universally admitted canon in scientific study that it is more probable that a law which is known to be without exception in phenomena, which we can clearly trace, extends to similar phenomena not yet fully explained, rather than that a new law should now come into play. This, of course, does not exclude the possibility of a new law, and the true scientist will cheerfully accept such a law, whenever by observation, comparison and experiment its correctness is established.

2d Epoch (continued). The other phase of the epoch under consideration relates solely to the origin of infusorial life. The microscope had been of great service in enabling scientists to account for the mode of generation in known animals, but with all this extension of knowledge it had also brought into view a new outlying territory which swarmed with animal life in numbers and kind before unsuspected. These are the infusoria—first discovered by Leeuwenhoek in 1675 and called by him anima-

---

migration of parasites—a subject just then being established—he took with him from Louvain to Paris four pups which he had reared. Two of them he had fed upon the *cysticercus cellulosus* of the rabbit, the larval form of the *tænia serrata* of the dog. These pups he presented to a commission of scientists (Valenciennes, Milne Edwards, Quatrefages and Jules Haime) saying: In two of these dogs you will find not a single specimen of *tænia serrata,* in the other two you will find many; and furthermore, in this one you will find specimens in four different stages of development, while in that one you will find them only in three stages, and the number of specimens in this dog is much greater than in that one. The pups were then killed and his statements were proved to be absolutely correct. In one dog, however, some *tæniae cucumerinae* were found, and Van Beneden frankly owned he could not tell where they came from. Since then it has been discovered that they originate from an acarus, the *trichodectes,* which lives in the hair of dogs and which is infested by the scolex of this variety of tape-worm. The dog licking its hair swallows the acarus and thus infects itself very much in the manner in which a horse is infected with bots, by licking up the eggs of the *œstrus* or gad fly.

cules. In 1764 Wiesberg gave them the name which they now
bear; this designation was made from the fact that they are always found in stagnant water and in infusions of both animal and
vegetable materials after short standing. Subsequently they were
studied by many observers, but it is to Ehrenberg* and Dujar-
din** that we are indebted for the most systematic description of
them, and their great works figure hundreds of varieties. The
almost illimitable numbers, the great diversity of form and of
organization, as well as the combined bulk of these microscopic
beings are almost beyond conception. We now know that there
are geological deposits of great size in different portions of the
earth's crust that consist almost exclusively of the calcareous
and silicious shells of these minute beings. Indeed Ehrenberg
himself regarded them " as forming by far the greatest number
and perhaps also the largest mass of living animal organisms on
the surface of the globe"† The rapidity of their development
is something wonderful, and being also infinitesimal in size their
mode of procreation was beyond the reach of the microscopes of the
day, and it is no surprise to learn that the old doctrine of spontaneous generation was again invoked to account for their origin.
This was done in 1748 by Needham††† and Buffon, who " led
by certain theoretical considerations doubted the applicability of
Redi's hypothesis to the infusorial animalcules, and Needham
endeavored to bring the question to an experimental test."††
Taking the juices of meats which had been extracted at a high
temperature he enclosed them in glass vials, also previously heated, corked them tightly and set them aside to cool. After a few

---

* *Die Infusionsthierchen, als vollkommene Organismen.* Leipzig, 1858.

** *Histoire Naturelle des Zoophytes Infusoires.* Paris,, 1841.

† DALTON: *loc. cit.* p. 127.

†† HUXLEY: *loc cit.*

††† JOHN TUBERVILLE: an English naturalist, 1713-1781. He was educated in the Roman Catholic faith and became a priest in that church, his life
being mostly spent on the Continent. He was Director of the Academy of
Maria Theresa, at Brussels. He devoted himself to scientific investigations
in connection with his work in teaching, and published many papers. His
principle work was a Treatise on Generation, published in French, the year
previous to his death.

days he invariably found in the vials infusoria present in great and constantly increasing numbers. His argument then was that if they were produced from germs, the germs must exist either in the substance which had been boiled, the water in which it was boiled, or in the air enclosed in the vial. Now boiling destroys the vitality of all germs, hence no infusoria should be developed in his infusions. But they were invariably present in his vials and accordingly he assumed that they were generated by a reorganization of the dead animal matter. But as Huxley says, most eloquently, "the great tragedy of science—the slaying of a beautiful hypothesis by an ugly fact—which is so constantly being enacted under the eyes of philosophers, was played almost immediately for the benefit of Buffon and Needham."

The Abbe Spallanzini* thought that Needham's experiments had not been conducted with sufficient care and precision, as no account had been taken of the absolute temperature to which the flasks and infusions had been subjected, nor had the mouths of the flasks been absolutely closed from contact with the external air. He therefore took glass flasks partly filled with organic infusions, and after closing them by hermetically sealing up the necks, exposed them to the temperature of boiling water† for an

---

* LAZZARO : an Italian physician, born in the duchy of Modena, 1729, died at Pavia, 1799. He was educated at Bologna, where he subsequently became a Professor, and still later he was appointed to the Chair of Natural History in the University at Pavia. He was one of the most eminent men of his day, an honorary member of nearly all the learned societies of Europe, and universally held in the highest esteem. His scientific studies were principally in physiology, especially of the lower animals; and by these studies which have been incorporated into the text-books, his name is more familiar to the medical student of to-day than that of many other recent observers. He refused offers of Professorships in a number of the prominent institutions of the time, among them the *Jardin des Plantes* in Paris.

† Various expedients for ridding the flasks of any existing infusoria or germs have been adopted by different experimenters. Those usually employed are

1. *Calcination*—causing air to pass through red-hot tubes.
2. *Filtration*—passing air through any substance which shall catch and retain all foreign matters.
3. *Subsidence*—allowing the particles to settle by gravity.
4. *Expulsion*—driving out air and particles contained therein by hermetically sealing neck of flask while contents are in an active state of ebullition.

Two or even more of these methods may of course be combined in a single experiment. The great difficulty of excluding all germs from the flasks by

hour. Then setting them aside at ordinary temperatures, which are favorable to the generation of infusoria, even after the lapse of months not a trace of animal life could be found on breaking the flasks. This was in the year 1775. But Needham was not satisfied with these results, and with a show of reason claimed that such a prolonged boiling would destroy not only germs, but the germinative, or as he called it "vegetative force" of the infusion itself. Spallanzini easily disposed of this objection by showing that when the infusions were again exposed to the air, no matter how severe or prolonged the boiling to which they had been subjected, the infusoria reappeared. His experiments were made in great numbers, with different infusions, and were conducted with the utmost care and precision. The result seemed convincing and was in substance that whenever animalcules were found in infusions which had been exposed to great heat, they " are not produced there because their germs have resisted this temperature or because they have been generated spontaneously ; but because new germs have been introduced into the infusion from the atmosphere after the boiling has ceased."

The naturalists of this period almost without exception acceded to these conclusions, doubt being entertained on a single point only. Oxygen had been discovered by Priestley in 1774, and its relation to the maintenance of life was for many years carefully studied by the physiologists. Now, might not the oxygen in the air of the flasks have been in some way altered by the high temperatures, and might not a renewal of oxygen be necessary to the development of life under any circumstances ? This certainly seemed reasonable, and so it became necessary to repeat the experiments under conditions which would obviate these objections. This was done in 1836 and 1837 by Schultze and

---

any method—even on the assumption that all those preëxisting within have been destroyed—may be appreciated when we recall the extremely minute size even of the fully developed parent animal. The *monas crepusculum* for instance is so small that eight millions of them would occupy a space no larger than a grain of mustard seed, and Prof. Owen has calculated that a single drop of water may contain five hundred millions of them. Such organisms would pass readily through imperceptible cracks or pores with changes in the temperature of the surrounding atmosphere.

Schwann, respectively. The first experimenter arranged his flasks with tubes bent at right angles and sealed to the stopper. To these tubes were attached a series of bulbs, which contained on one side anhydrous sulphuric acid and on the other a strong solution of caustic potash. Air was then by suction daily drawn into the flasks, passing in through the acid and emerging from the potash side. This process was continued for months (May to September) and no trace of infusoria, confervæ or fungi was found in the fluids. Schwann's experiments varied from Schultze's in that he passed air in through a series of bent tubes, which were heated up to 600° F. The results, however, were the same, and in both cases it was proven that the air or oxygen had undergone no change. Thus it seemed clear that whenever life made its appearance in the infusions in closed flasks it was produced by germs introduced from without, and in the experiments under consideration, the germs in the atmosphere (if there were any) were destroyed by the acid and the calcination.

These experiments were accepted almost universally as demonstrative of the incorrectness of the theory of spontaneous generation and it may be said with propriety that within a few years subsequently the question was deemed to have been put to rest for all time. But a rigorous analysis of the evidence shows, as Prof. Huxley has very justly pointed out, that this conclusion is not warrantable. All that the experiments really proved was " that the treatment to which the contents of the flasks had been submitted had destroyed something that was essential to the development of life. This something might be solid, fluid or gaseous ; that it consisted of germs remained only a hypothesis of more or less probability," and, no one, it must be remembered, had ever yet seen the germs. Helmholtz, in 1843, by his experiments narrowed this issue by showing that the interposition of a membrane between a putrefying (swarming with life) solution and one that is simple putrescible prevents the development of organisms in the latter. The cause of the development must therefore be something that cannot pass through the membrane. But gases and fluids can readily pass through, and hence it follows that it must be either a colloid or solid matter. Next in point of order Drs. Schroeder and Von Dusch* helped clear up up this point by showing that the simple exclusion of air from an infusion by a plug of cotton wool prevented both fermentation and development of organisms; and finally Tyndall settled the matter definitely by showing that ordinary air is full of solid particles of matter, and that they are entirely strained out by filtration through the plug of wool. It only remains, therefore, to prove that among these particles are germs which, under appropriate conditions, are capable of being developed into animal life. "This," says Huxley, "was done by M. Pasteur, in those beautiful researches† which will ever render his name famous, and which, in spite of all attacks upon them, appear to me to be models of accurate experimentation and logical reasoning." In point of time, however, this demonstration was not made until the third or last epoch in the history of spontaneous generation.

3d Epoch. This dates from the year 1858, when the question—which by general consent had been considered as closed—was reopened Paris by Pouchet, the distinguished naturalist of Rouen. He sent a communication to the French Academy in which he declared that he had experimentally proven the truth of spontaneous generation, and in the following year he published his well-known work on the subject.‡ It may be well to note just here, as bearing upon the reliability of his evidence and arguments, that he was undoubtedly influenced by a motive, for in a preface by him to Pennetier's work on the origin of life, he says: "For all reflecting minds heterogenous production is a logical consequence of the appearance and ascending development of organized beings upon the globe."§ Furthermore, in the very opening paragraph of the preface to his own book, he uses this expression: "When *by meditation* it was evident to me that spontaneous generation was one of the means employed by

---

* *Annal. de Chimie,* tome XLI., 1851, and *Chemical News,* Vol. V., 1862.

† Prof. Tyndall, somewhat enthusiastically says, that his "labors in connexion with this subject may be fitly called immortal."—*Lancet,* February 12, 1876, p. 262.

‡ *Hétérogénie; ou Traité de la Génération spontanée, basé sur des Nouvelles Expériences.* Paris, 1859.

§ Quoted by Dalton, *loc. cit.*

matter for the production of living beings, etc." This motive it is easy to see grew out of the tendency of the geological studies of the day, which show that the earliest remains of animal forms found in the earth's crust belong to the lower orders and gradually ascend in successive epochs to man., Hence it was an easy —I do not say legitimate—inference, that the higher orders had been gradually evolved out of the lower. And as the deepest and oldest geological strata show no organic remains it is a fair assumption that at some time in the great past there was no life on the globe ; and hence another easy inference, that the first living beings which appeared were produced by spontaneous generation. This is the gist of the evolution theory and as an inducing motive in Pouchet's advocacy of spontaneous generation it is worthy of remembrance ; for let me again remind you that such a question cannot in the nature of things be argued on the ground of probability, but must be determined solely by experimental evidence.

Pouchet further asserted that he had repeated Schultze's experiments with every possible precaution, but with totally different results. These assertions attracted much attention, and a few scientists sided with him, though the majority and among them the most of the leading physiologists opposed him. The question assumed different phases, and in January, 1860, it was made one of the prizes of the Academy. Pasteur, at this point, took up the matter and made the researches to which allusion has already been made. His first step was to ascertain whether in reality there are floating in the atmosphere spores of the microscopic fungi or germs of the infusoria, for by this time the question was confined almost exclusively to one point. viz : the atmosphere as the supposed source of the organic germs. For this purpose he passed air through a wad of gun cotton packed in a glass tube. Then dissolving the cotton in ether and alcohol he was enabled to gather in the deposit whatever floating particles had been caught upon the cotton during the forced passage of the air. Then examining these deposits with the microscope he found, besides the easily recognizable matters such as starch-granules, hairs, coal-dust, etc., which are known to be floating in the atmosphere, numerous round or oval organized corpuscles, some of which "closely resemble the spores of the commonest moulds," and others "resemble the globular infusoria and are regarded as being the eggs of these small beings." " But, as to affirming," he says " that this particular one is a spore, or still more that it is a spore of a definite species, or that that corpuscle is the egg of an infusoria or of such a species, I do not believe that this is possible. I am content, as far as I am concerned, to affirm that these corpuscles are evidently organized." In the light of Lemaire's subsequent observations, which will soon be alluded to, it is demonstrated that this opinion of Pasteur's was correct.

Now assuming that such germs are floating in the atmosphere, Pasteur asserted that their number and variety would differ greatly in given volumes of air collected from different localities, and he even said in definite terms "that everywhere it was possible to detach a volume of air from the atmosphere which will contain neither egg nor spore, and will not produce generation in putrescible solutions." To determine this point he prepared a large number of flasks partly filled with solutions of sugar and yeast. After thorough boiling the flasks were hermetically sealed by drawing out the necks to a fine point. The flasks were then taken to different localities and opened by pinching off the necks. Air would rush in by reason of the partial vacuum which had been formed by the boiling of the contained fluid, and thus air from any locality could be gathered for experiment. In this way air was taken from the tops of high mountains, in the very midst of glaciers, from level, open plains in the country, from the streets of crowded cities, from cellars, etc. The result was that just in proportion to the distance from crowded cities, and the absence of disturbing currents in the atmosphere, the evidences of organic life diminished. Of twenty flasks which were opened on the " *Mer de Glace*," in the Alps, at an altitude of 6,000 feet, one only subsequently contained any trace of life. In another series of experiments flasks were filled in the cellars of the Observatory in Paris, where the temperature is almost uniform and the air is very still. The number containing organisms was very much less than in those filled in the garden of the same building. Pasteur predicted that if flasks could be opened and closed in deep cellars with absolutely no disturbance caused by the entrance of the operator there would be the same absence of vitality as in flasks which had been long exposed to red heat.* Later on he learned, what he had also predicted, that by simply turning the long neck of his flasks† downward they might be kept indefinitely without sealing or stoppers of cotton, and still no organisms would show themselves, for by this simple expedient the germs could not enter the flasks, gravity opposing.

On the other hand, Pouchet, assisted by MM. Joly and Musset, shortly afterward went over the same ground. He collected the solid particles from the air by means of an instrument which he called an aeroscope. This was a simple tube drawn out to a point. Air was passed in a jet through this and made to impinge upon a glass plate covered with some viscous substance. A pile of dust was thus caught, and then submitted to examination by the microscope. But strangely enough although he found plenty of foreign materials, like coal-dust, starch-granules, etc., not a trace of organic life, either in shape of spores or of germs. He was untiring in his researches. "He has," says M. Joly, "examined the dust which finds its way into the respiratory cavities in man and the lower animals; that which has been the accumulation of centuries in our Gothic cathedrals, and that which floats in the air of our public halls, our theatres, and our hospitals. He has crossed seas, climbed high moun-

---

* In an unexpected direction Prof. Tyndall by his recent experiments with closed boxes has practically verified this prediction, though the experiments were not made with cellar air. The similarity of the two cases, however, is apparent. *Vide postea*, p. 175.

† Many of Pasteur's flasks are still preserved in Paris, and by repeated examination have been found to remain unchanged. As late as November, 1874, M. Balard, in presenting to the Academy of Sciences a paper by M. Servel, detailing experiments which the writer held to be demonstrative of spontaneous generation, took occasion to say that he had just then examined in Pasteur's laboratory some of his *unsealed* flasks which contained blood that had been drawn more than eleven years previously, and in which, during all this time, no bacteria had appeared and no putrefaction had taken place.

tains, descended into the crater of Vesuvius and of Etna; he has penetrated even into the tombs of the Pharaohs and studied their crania blackened and dusty with the lapse of time."* It seems too bad after all this that he should not have been rewarded by occasionally finding germs, but his results were barren. Then he prepared a series of flasks with putrescible infusions as Pasteur had done and with the same end in view, of gathering air from different localities and learning whether subsequently organisms would develope in them. Now, holding as he did to the theory of spontaneous generation, he said, as the chemical constitution of the air is the same everywhere, we ought always to find such organisms wherever air, no matter from what part of the globe it may be taken, is brought into contact with putrescible solutions. And sure enough his flasks were found always to contain them.

Here then was a flat contradiction in the results obtained in each series of experiments by these two eminent observers. Each showed, at least to his own satisfaction, the fallacies in the experiments of the other, but the possibility of reconciliation seemed almost hopeless. It was therefore proposed to submit the case to a jury of experts to be selected by the Academy. The contestants availed themselves of this proposal, and a commission consisting of Flourens, Dumas, Brongiart, Milne Edwards and Balard was appointed in January, but did not begin its labors until June, 1864. Each contestant stated his propositions in definite and unmistakeable terms, and M. Joly, in his confidence, even went to the extent of saying, "If a single one of our flasks remains unchanged, we will acknowledge defeat." Pasteur appeared with sixty flasks and made his experiments. Pouchet and his confreres then declared that they were unwilling to abide by a decision on this series of experiments, and as the commission persisted in holding both sides to this series which

---

* *Les Générations spontanées*: Par JULES JAMIN, *Revue des deux Mondes*, Vol LIV., p. 131. Though somewhat argumentative, this article is a good resumé of the controversy before the French Academy, and the results on both sides are clearly set forth. A translation of the article was published in the *Methodist Quarterly Review* of Oct., 1865.

had been the principal cause of the controversy, Pouchet withdrew from the contest.* The Commission, however, continued its investigations, and in February of the next year they reported that the facts which were observed by M. Pasteur and contested by MM. Pouchet, Joly and Musset were of the most perfect exactitude.**

One point only needs now to be settled in order to complete the chain of evidence and render demonstration complete. Fortunately this was done before the rendition of the report. That is to collect and identify the germs from the atmosphere, and to propagate them, for it will be remembered that with a commendable prudence Pasteur had only stated his opinions as to the corpuscles and spores, which he had gathered in the manner already described. The heterogenists with force and with reason said, if such organic bodies are floating in the atmosphere, it is only fair that our opponents should show them to us. This was accomplished by Dr. Lemaire and Prof. Gratiolet in 1864. They condensed the moisture of the atmosphere in a wide open vessel,

---

* "It is, perhaps, unfortunate," says Jules Jamin, "that the Commission held so stringently to the programme as to let slip the unique opportunity of a solution which was expected from it. But it is evidently clear that the heterogenists, however they may have colored their retreat, were self-condemned. If they had been sure of the fact which they had solemnly undertaken to prove under penalty of acknowledgment of defeat, they would have persisted in proving it, for it would have been the triumph of their doctrine. It is doubtful causes only that are allowed to go by default."—*Revue des deux Mondes*, Vol. LIV, p. 438.

** En resumé, les faits observeés par M. Pasteur et contestés par MM Pouchet, Joly et Musset, sont de la plus parfaite exactitude. Des liquers fermentescibles peuvent rester, soit au contact de l'air confiné, soit au contact de l'air souvent renouvelé, sans s'altérer, et quand sous l'influence de ce fluide il s'y développe des organismes vivants, ce n'est pas à ses éléments gaseux qúil faut attribuer ce développment mais à des particules solides dont on peut dépouiller par les moyens divers, ainsi que M. Pasteur l'avait affirmé. *Comptes Rendus*, Vol. lx, p. 396.

Running through the *Comptes Rendus* from the year 1858 to 1865, the reader will find all the facts and reports upon this remarkable controversy, which, as Dr. Dalton remarks, may almost be said to have kept the Academy in a turmoil for some six or seven years, and which at times was so conducted as to provoke considerable bad feeling.

A good sketch also of Pasteur's experiments may be found in Schützenberger on Fermentations, Vol. xx, of the International Scientific Series, published by D. Appleton & Co., New York.

which was surrounded by ice. Water was thus obtained from
different localities. It was carefully enclosed in glass tubes and
submitted to examination. The liquid thus condensed was at
first, colorless, clear, and contained no living being. There
were, however, "myriads of spherical roundish and fusiform
spores, pale cells and semi-transparent ovoid bodies," besides, of
course, the foreign matters. At the end of fifteen hours large
numbers of living bacteria were found; in forty-eight hours vibrios and spirilla swarmed in abundance, and in three days
monads, whose incubation is slower, were also present. *Just in
proportion as this mass of life appeared, the spores and semi-transparent corpuscles disappeared."* These experiments, varied in
many ways, even to the extent of sowing as seed the particles obtained from the air, and thus propagating infusorial life, were
demonstrative of the actual existence of organic germs in the
atmosphere. In this way the work of the Commission was materially aided, and the decision which they rendered was generally
accepted as conclusive against spontaneous generation.

The heterogenists, however, even to-day do not accept these
conclusions, and, although they grant that the usual mode of the
development of infusorial life is from pre-existing germs, they
claim that, under exceptional conditions, it may arise spontaneously. Foremost among these advocates, and conspicuous for
his attainments, is Dr. H. Charlton Bastian, of London. His
labors in this direction and his publications, both fugitive and
systematic,[1] are familiar to you all. I shall, therefore, for lack
of time attempt no description, not even a summary of his experiments, but this sketch would be incomplete without a reference to them. I would in no way underestimate their importance
as contributions to our knowledge on the subject, but after examination of all the evidence which has been accessible to me, I
am unable to see that his work has advanced the main question

1 *The Modes of the Origin of Lowest Organisms, including a discussion of the
Experiments of M. Pasteur, and a Reply to some Statements by Profs. Huxley and
Tyndall.* London, 1871.

*The Beginnings of Life: being some Account of the Nature, Modes of Origin
and Transformations of Lower Organisms.* 2 Vols., London, 1872.

materially beyond the point where the French Academy left it. In some minor particulars, his work has been of great service. His experiments have been analysed very carefully, and in many cases repeated by Frankland,[1] Burdon Sanderson,[2] Sedgwick, Lionel Beale, Roberts,[3] Lankester,[4] Tyndall,[5] Huxley[6] and others, and numerous sources of error of pointed out. Some of these authorities are believers in the possibility of spontaneous generation and later, on p. 170, I have quoted opinions from them, but no one of them, so far as I am aware, admits that Dr. Bastian's experiments are conclusive, or that spontaneous generation has ever yet been demonstrated. Some of them, too, have been very wrongly quoted by Dr. B. and other heterogenists as supporting their views. On the other hand there are still some who openly avow their belief in the doctrine, and I may mention here a few names that occur to me at this writing of men eminent for their scientific attainments—Mr. Wallace,[7] Prof. Huizinger,[8] of the University of Gröningen, Prof. Cantoni and others, of the University of Pavia, and Ernst Haeckel.[9] The latter, however, is a

---

(1) *Nature*, Vol. III, p. 225.

(2) *Ibidem*, Vol. IV, p. 377, and Vol. VIII, pp. 111, 181.

(3) Mr. Roberts was forced, by his own experiments, apparently very unwillingly, to believe in the possibility of spontaneous generation. Prof. Tyndall, in his recent paper before the Royal Society, (p. 174), has pointed out a minute error of detail which vitiated the results and led Mr. Roberts to his conclusions. For his experiments see *Philosphical Transactions* Vol. CLXIV., 1874.

(4) *Nature*, January 30, 1873, page 212, and Oct. 16, 1873, p. 505. Lankester and Podes original experiments are reported in detail in *Proceedings Royal Soc.*, Vol. XXI., 1873.

(5) *Medical Times and Gazette*, Oct., 1870, page 406, and *The Lancet*, February 12, 1876, p. 262.

(6) *Nature*, Vol. II., p. 473.

(7) An elaborate review of Dr. Bastian's larger work was published by Mr. Wallace, in *Nature*, Aug. 8 and 15, 1872.

(8) *Pflüger's Archiv*, Vol. VII., p. 549. An advance summary by himself of the experiments detailed in this paper may be found in *Nature*, March 20, 1873, p. 380. See also comments on the same by J. Burdon Sanderson. *Ibidem*, Oct. 2, 1873, p. 178, and *Med. Times and Gazette*, Sept. 27, 1873, p. 341. Sanderson interprets these experiments differently from Dr. Bastian and Huizinger himself.

(9) Prof. E. Ray Lankester has published a lengthy abstract of Haeckel's opinions in *Nature*, March 2, 1871, p. 355.

supporter of this side of the case from purely theoretical considerations, for, although he concedes that, thus far there is no absolute proof of the theory, he holds that the difficulties in the
way of ultimately establishing it are not only surmountable but
less formidable than those that face the supporters of Biogenesis.

I may sum up then, without further detail, the whole matter,
by saying that there is no trustworthy evidence to-day that spontaneous generation has been demonstrated in a single instance.
Even Huxley, who declares that if it were given him " to look
beyond the abyss of geologically recorded time to the still more re-
mote period when the earth was passing through physical and
chemical conditions which it can no more see again than a man
can recall his infancy, he should expect to be a witness of the
evolution of living protaplasm from not living matter" says very
significantly that, with this limitation, Redi's great doctrine of
biogenesis seems to him victorious along the whole line at the
present day. Authority*, however, cannot settle the question,

---

* Notwithstanding this admission, I venture to add here a few opinions
which I have noted, without special search, in the course of my reading.
Even if not decisive of the question at issue, such opinions are interesting.

SIR WM. THOMPSON: "I am ready to adopt as an article of scientific faith,
true through all space and through all time, that life proceeds from life and
from nothing else." President's Address, British Association, 1874. *Nature*,
Vol. IV. p. 269.

HAECKEL: "Positive contradiction of the hypothesis of Archigenesis is
impossible. Positive proof there is not yet since no one has yet seen any organism take origin except by parentage. * * * Either the *monera* were
once for all at the beginning of organic life on the earth produced by Archigenesis, * * *' or in the course of the earth's history they have been produced by recurring acts of Archigenesis, and in this case there is no reason
why this process should not occur at the present time." *Nature*, March 2,
1871, p. 356.

J. BURDON SANDERSON: "I do not hold that spontaneous generation is
impossible. I do not regard heterogenists as scientific heretics. All I say is,
that up to the present moment I am not aware of any proof that they are
right." *Nature*, Oct. 2, 1873, p. 179.

M. FLOURENS: "So long as my opinion was not formed, I had nothing to
say. Now that it is formed I will express it. Pasteur's experiments are decisive. If spontaneous generation is a reality what is necessary to produce animalcules? Air and putrescible fluids. Now M. Pasteur puts air and putrescible liquids together and nothing comes of it. There is then no spontaneous

and in entering this Scotch verdict of *not proven*, I simply accept what I believe to be the correct interpretation of the best attainable evidence in the present state of our science. If new evidence can be adduced which is subversive of this conclusion, we must accept it without regard to our predilections or beliefs. To reject the theory on such considerations is contrary to the scientific method, and it is by this method alone that experimental evidence should be interpreted.

---

generation. To doubt longer is not to comprehend the question."—*Revue des deux Mondes*, Vol. LIV, 1864, p. 141.

PASTEUR; "This conclusion which I have already formulated is unassailable. *In the present state of science the hypothesis of spontaneous generation is a chimera.*" Translated from a letter to Prof. Tyndall, dated Paris, February 8, 1876. *The Lancet*, Feb. 19, 1876, p. 296.

HUXLEY; "If in the present state of science, the alternative is offered us, either germs can stand a greater heat than has been supposed, or the molecules of dead matter, for no valid or intelligible reason that is assigned, are able to rearrange themselves into living bodies, exactly such as can be demonstrated to be frequently produced in another way, I cannot understand how choice can be, even for a moment doubtful. But though I cannot express this conviction of mine too strongly, I must carefully guard myself against the supposition that I intend to suggest that no such thing as Abiogenesis ever has taken place in the past or ever will take place in the future. * * * All I feel justified in affirming is that I see no reason for believing that the feat has been performed yet."—President's Address, British Association, 1870. *Nature*, Sept. 15, 1870, p. 403.

BASTIAN: "On account of this *à priori* probability and in the face of this evidence, I am, therefore, content, and, as I think, justified in believing that Living things may and do arise *de novo*." Dr. B's. views have been often summarized in his prolific writings, but I know of no more concise expression of them than the above, from a closing paragraph of a series of papers "on the Heterogeneous Evolution of Living Things."—*Nature*, July 14, 1870, p. 228.

LIONEL S. BEALE: "I confess to being an opponent of the doctrine, but simply because I cannot admit that the evidence yet adduced is at all convincing. * * * The fact of *à priori* arguments having been so very much dwelt upon, makes me think that the mind of the experimenter may have been to some extent prejudiced (prepossessed) in favor of the doctrine he seeks to support by new facts, and in this way they are calculated to excite in my mind, however much I may resist, a doubt whether the inferences which have been arrived at really have been deduced from facts of observation and experiment *only*."—*Nature*, July 28, 1870, p. 254.

SMITH, WORTHINGTON G,: "It seems to me rational enough to suppose that unicellular bodies and objects of the lowest possible organization may be heterogeneously produced from the inorganic world."—*Nature*, August 4, 1870, p. 276.

VALENTIN: Prof. Physiology, Univ. Bern. "On the whole, the hypothesis of a spontaneous generation of plants or animals can only be regarded as a

An important, if indeed it be not the pivotal point in the recent discussions of this question, is the degree of heat to which vegetable and animal germs can be submitted without destroying their vitality. On this point there is great discrepancy of opinion. To admit that $212°$ F. is insufficient, is to destroy absolutely, as Pouchet pointed out, the validity of Spallanzini's and Schultze's experiments. I allude to this, not for argument's sake, which is foreign to a historical review of the question, but merely to enable me to state several interesting facts. Prof. Jeffries Wyman*, of Harvard College, found infusoria in infusions that had been boiled four hours, but he found none after five or six hours boiling.

kind of superstition which is constantly receding before the advance of the natural sciences."—*Text-Book of Physiology.* London, 1853, p. 624.

POUCHET, in his *Hétérogénie,* previously alluded to, (p. 162) deliberately quotes Valentin as a supporter of his (Pouchet's) views, whereas he is an uncompromising opponent of them.

VAN BENEDEN: "It is evident to all those who place facts above hypotheses and prejudices, that spontaneous generation, * * * does not exist, at least if we only consider the present epoch."—*Animal Parasites and Messmates,* p. 106.

CARPENTER, W, B.: "The doctrine of 'spontaneous generation' cannot now be said to have any claim whatever to be received as even a possible hypothesis."—*Principles of Physiology, Genl. and Comp.,* 3d edition, Philadelphia, 1851, p. 866.

HUIZINGER: "Under the above described circumstances (i. e. his experiments) Bacteria can arise without pre-existing germs. Not in any single case have I seen any other organisms than Bacteria—never fungi."—*Nature,* March 30, 1873, p. 381.

JAMES SAMUELSON: "If the believers in spontaneous generation still insist that their hypothesis has not been refuted and that, assuming my observations to be correct, their view of the case has not been fully disproved, I am not prepared to deny this. But, on the other hand, I must be permitted to retort that their experiments have only proved, so far, their inability, notwithstanding all their precautions, to exclude invisible germs from their infusions."—*Med. Times* and *Gazette,* Sept. 24, 1870, p. 376.

TYNDALL: "As far as inquiry has hitherto penetrated, life has never been proved to appear independently of antecedent life."—*Nature,* February 3, 1876, page 269.

* PROF. WYMAN'S experiments are justly deemed among the most valuable contributions ever made to this subject. They are recorded in the *Amer. Jour. of Sci. and Arts,* Vol. XXXIV, 1862, p. 79, and Vol. XLIV, 1867, p. 152.

Professor Cantoni,* of the University of Pavia, by means of a Papin's digester, carried the temperature of his flasks to 110°– 117° Centigrade, and yet vibrios in large number, were produced in two days. And Dr. Bastian has repeatedly noted a temperature ranging from 148° to 150° C. (equal to 312 F.) to which his infusions have been submitted for a brief period, and yet living matter has soon developed. On the other hand, some of the living infusoria will flourish in temperatures below the freezing point. They may even be dried and kept for years and then by the application of moisture they will revive and resume active motion. The truth seems to be that the germs, at least, are of well nigh inextinguishable** vitality, but it is difficult, if not impossible, to secure consent as to the precise limits of temperature, wet and dry, and other conditions under which such vitality may be retained. Until these points can be determined it seems almost hopeless to expect a solution which will command assent from both parties to the controversy.

---

* *Gaz. Med. Ital. Lombard.*, Ser. VI., Vol. 1, 1868. It is a significant fact in this connection that Cantoni's, Wyman's, Huizinger's, and others' experiments have been interpreted by the adherents of opposite sides of the question as substantiating their own view of the matter.

** " The tenacity of life [of the Rotiferæ] is one of the most extraordinary phenomena. Their resistance to cold is something marvellous, and we don't even know where it stops; the lowest temperature that we can obtain in our laboratories does not seem to have any effect upon them * * * I have sometimes removed them quickly from the freezing apparatus and thrown them into a stove heated to 176° Fahr. * * * In this two-fold test and formidable transition from cold to heat, these microzoa passed rapidly through a change of 216° Fahr. without being in the least inconvenienced by it." Pouchet. *The Universe*, p. 56. 2d Ed., 1871.

Rudolphi long ago learned that the entozoa in frozen fish when thawed out resumed their customary activity. Frankland has recently described ice-fleas, which flourish in the glaciers of the Alps, at a temperature constantly below freezing point.

Mr. Bauer kept the *vibrio tritici*–a parasite of wheat–dried for seven years, and on moistening them with water they resumed their active motions.

M. Balbiani in 1857 "observed a drop of water on a plate of glass in which were living colpods. When the water was evaporated each became encysted and dormant in its envelope. The plate was moistened again in 1861, when every colpod was observed to come out from its shell and promptly resume its vital functions, which had been interrupted by seven years of sleep."—*Revue des deux Mondes*, Vol. LIV., 1861, p. 439.

Though not pertinent, strictly speaking, to a historical sketch like this of the doctrine of spontaneous generation, it is interesting to know what are the organisms whose germs can withstand such savage treatment as that described, and whose mode of generation is involved in any obscurity. They are only t̨ ; very lowest orders of life known, and by common consent the question is limited to the *Monas, Vibrio, Spirilla* and *Bactirium.* The last three are generally conceded to belong to the vegetable world. At all events, they stand upon the extremest limits of that debatable ground in which spontaneous generation, if it is ever shown to be a reality, must be found. The higher orders, comprising almost the entire class of infusoria, are now known to reproduce themselves by true sexual generation.*

The last, but by no means most unimportant contribution to this subject is by Prof. Tyndall; and I cannot close this sketch without alluding to his work, although it was not undertaken for the purpose of solving or attempting to solve the problem of spontaneous generation. His investigations† were a continuation, practically, of his former experiments on floating particles in the air (to which allusion has already been made, p. 162), and were for the purpose of supplying direct evidence to connect

---

* The observations of Stein, Englemann, Balbiani and others have clearly established this fact for a large share of the infusoria. Their labors have been supplemented and their results corroborated by the elaborate studies of Ernst Eberhard. An abstract of his work may be found in *Quar. Jour. Micros. Sci.* New Series, Vol, VIII, p. 155. See DALTON, *loc. cit.* for reference to Stein and others.

† *Medical Times and Gazette,* January 29, 1876. Also the *Lancet,* same date. The title of his paper is "On the Optical Deportment of the Atmosphere in Reference to the Phenomena of Putrefaction and Infection." An abstract by the author himself was published in *Nature,* February 3, 1876.

Since this lecture was given Dr. Bastian has sharply criticised both Tyndall's experiments and his deductions. In turn Prof. Tyndall has made a rejoinder of equal positiveness and severity. Dr. B. entitles Tyndall's paper "a new attempt to establish the truth of the germ theory," and then unsparingly attacks not the germ theory of disease but the doctrine of biogenesis—questions which are far from being identical. A most dispassionate review of Dr. Bastian's position in this controversy may be found in the *Popular Science Review,* 1876, in a paper by Rev. W. H. Dallinger, V.P.R.M.S. This paper contains also some valuable discoveries made by the author and Dr. Drysdale.

zymotic changes (putrefaction) with the presence of such particles. The experiments show that there are particles of matter
in the air which are ultra-microscopic in size, and yet their
presence can be made evident by their power of refracting light,
and *secondly*, whenever air deprived of such particles is brought
in contact with putrescible solutions no putrefaction occurs, ' *i*
that it invariably occurs wherever the air does contain such particles. An air tight box with a glass side and windows in the
ends had sealed to its bottom twelve test tubes with their mouths
upward and projecting inside the box. In the top was an India
rubber stuffing box through which passed a glass tube by means
of which infusions could be dropped into the test tubes. The
whole interior was smeared with glycerine and the apparatus
allowed to stand a number of days. By subsidence, then, all
foreign matter was caught and retained on the bottom. Now
when the electric beam is passed through the box, the space
inside appears perfectly dark, and the freedom from dust is thus
proven, for, so long as there are floating particles, the track of
the beam can be detected. Then organic solutions of different
sorts were dropped into the tubes and boiled from below (the
tubes being made to project from the bottom of the box, for this
purpose, for a space only of five minutes. In no single instance,
except where the cause of the failure was obvious, did any turbid-
ity occur in the solution, nor was organic life (bacteria) found
after even a lapse of weeks and of months. The conclusion thus
reached by Prof. Tyndall is that the power of scattering light
and of developing bacterial life by the atmosphere go hand in
hand, and both are dependent upon the presence of particles
which, even by the highest powers of the microscope, we cannot
detect. As bacterial life is regarded as necessary to putrefaction,
incidentally, therefore, spontaneous generation is held to be neg-
atived by these experiments. This deduction will be accepted or
rejected according as one is inclined to side with one or the other
parties to the controversy. I cannot refrain from expressing my
own opinion, that not only are the experiments of the utmost
exactness, but also that already they have led to results which
will contribute very materially to the ultimate solution of this
problem.

M.

EXPERIMENT TO SHOW THAT CHLOROPHYLL BODIES MIGRATE UNDER THE INFLUENCE OF VARYING INTENSITIES OF LIGHT.

BY V. M. SPALDING.

If we make a thin section of a green leaf from a tree or flowering plant, or, better still, if we select a fresh, delicate leaf from a clump of growing moss, and subject it to examination with the microscope, it will be seen that, like nearly all other plant tissue, the leaf is made up of distinct cells. To consider a special case at once, it will be well to notice the structure of the leaf employed in the experiment, which belonged to a common species of moss, *Mnium affine.*

The leaf is very small, not more than one-fifth of an inch long and about two-thirds as wide. It is very delicate, almost transparent, and can be examined under the microscope without any preparation whatever. Thus examined, its structure is seen to be exceedingly simple. It is made up of a single layer of cells, except along the midrib, where it is two or three layers thick. No vascular bundles are found, and no stomates. We have, therefore, scarcely more to examine thon a single layer of cells with their contents, so that the experiment can be performed with the utmost readiness and certainty.

If now one of these cells is examined, it will be found to consist of a delicate, membranous cell-wall, containing a transparent, fluid-like substance, in which float from twelve to twenty round, green bodies. These latter are the so-called chlorophyll bodies.

We have nothing to do just now with the question how these bodies came there and what they are for, though it suggests a very interesting line of investigation. The position of the chlorophyll bodies in the cell, and their movements under the influence of varying intensities of light, are all that will here be considered.

If the plant has been kept on a short allowance of light, the chlorophyll bodies will be found arranged in planes parallel to the leaf surface, as if they would expose themselves as fully as possible to the little light which falls upon them ; but if the plant has been in bright sunlight, they will be found in line along the side-walls of each cell, as if to hide themselves in a measure from the intense light. And further, if the plant which has been in darkness or in diffused light, and in whose cells the chlorophyll bodies are arranged in the former position, is now placed in direct sunlight, it will take but a short time for them to assume the latter position, and *vice versa.*

This curious fact has been known only a few years, but within that time it has been repeatedly observed by the best European botanists, and is now as well established as any other phenomenon of vegetable life.

With a view to confirming this fact by actual observation, a detailed experiment was performed, March 4th, 1876. A bunch of moss had been kept in a room for several days. No special pains were taken with it, except that it was kept moist and for two days had been turned bottom upwards so as to cut off the light more effectually. It was in good, healthy condition, with soil around the roots which had been taken up with it when it was brought from the woods. It had not, then, been exposed to direct sunlight for several days, and if any light reached it at all, it must have been very faint.

Just before noon, six fresh leaves were examined, from different stems of the bunch. In this and in the subsequent examinations, pains were taken to select delicate young leaves, it being supposed that they would be most susceptible to the influence to be tested. The result was as follows : In *five* out of the six leaves, the chlorophyll bodies were found to be distributed, in all the cells, in planes parallel to the surface of the leaf. In the remaining leaf, this was true except in the cells near the base, where they were arranged perpendicular to the surface, and in a line around the side-walls.

The plant was then again moistened with fresh water, and hung in a window where it was exposed to the action of direct sunlight. At the end of an hour, two leaves were examined and no change was detected. At the end of three hours, the plant having been in bright sunlight all the time, six leaves were examined, those being selected which had been certainly exposed to the light, but not withered by it. The following result was obtained: In *five* leaves out of the six, the chlorophyll bodies were arranged in circular lines around the side-walls of the cells, perpendicular to the leaf surface. In the sixth leaf, the same was true for the cells near the apex, while in those near the base the chlorophyll bodies were arranged in planes parallel to the surface. Nothing could be more distinct than the green chlorophyll bodies seen in this last examination ; they could be readily counted, and so perfectly and uniformly were they arranged around the side-walls of the cells that in many cases not a single one stood out from the line.

It will be remembered that the moss-leaves already exam ined had been kept in darkness for some time previous to the experiment. It now remains to notice the position assumed by the chlorophyll bodies in the presence of diffused light. On the same day, about noon, six leaves were selected from a part of the same bunch of moss that had been kept for two days near an east window, and which had, therefore, been exposed most of the day time to diffused light. The results were nearly identical with those obtained in the examination of the leaves which were kept in darkness, though in one leaf the chlorophyll bodies were throughout arranged about the side-walls.

The obvious conclusion, then, must be that the chlorophyll bodies when deprived of the strong light of the sun, whether they are in darkness or in ordinary diffused light, arrange themselves in planes parallel to the surface of the leaf, thus exposing as much of their surface as possible to the action of light, but when exposed to the action of direct sunlight, arrange themselves around the side-walls of the cells, thus presenting a less amount of their surface to the intense light.

It may be added that at the date of writing, March 29th, over three weeks after the observations above recorded were made, an examination was made of the leaves of the same moss, which had been left in a faintly lighted closet since the former date, without care of any kind. All the leaves examined were found to have their chlorophyll grains arranged in planes parallel to the leaf surface, thus showing that after removal from the sunlight these bodies had again arranged themselves in their former position.

If $\frac{1}{200}$ of an inch be taken as the diameter of an ordinary cell—and this is a rather large measurement for this species—it will be seen that chlorophyll bodies pass through a very small space in performing their migrations to and from the cell-walls. Supposing one of the chlorophyll bodies to be exactly in the middle of the cell, it cannot pass through more than $\frac{1}{400}$ of an inch before reaching the cell-walls. At this rate, then, about a year will be consumed in traveling a single inch. Still it is pleasant to think that the mites who chance to be born in a moss-cell have traveling facilities which, though a trifle slow, are, nevertheless, safe and regular.

www.ingramcontent.com/pod-product-compliance
Lightning Source LLC
Chambersburg PA
CBHW021804190326
41518CB00007B/443